T0220230

HOW TO SPEAK TECH

THE NON-TECHIE'S GUIDE TO KEY TECHNOLOGY CONCEPTS

SECOND EDITION

Vinay Trivedi

Apress®

How to Speak Tech: The Non-Techie's Guide to Key Technology Concepts

Vinay Trivedi
Newtown, PA
USA

ISBN-13 (pbk): 978-1-4842-4323-7   ISBN-13 (electronic): 978-1-4842-4324-4
https://doi.org/10.1007/978-1-4842-4324-4

Library of Congress Control Number: 2019934987

Managing Director, Apress Media LLC: Welmoed Spahr
Acquisitions Editor: Shiva Ramachandran
Development Editor: Laura Berendson
Coordinating Editor: Rita Fernando

Cover designed by eStudioCalamar

Distributed to the book trade worldwide by Springer Science+Business Media New York, 233 Spring Street, 6th Floor, New York, NY 10013. Phone 1-800-SPRINGER, fax (201) 348-4505, e-mail orders-ny@springer-sbm.com, or visit www.springeronline.com. Apress Media, LLC is a California LLC and the sole member (owner) is Springer Science + Business Media Finance Inc (SSBM Finance Inc). SSBM Finance Inc is a **Delaware** corporation.

For information on translations, please e-mail rights@apress.com, or visit http://www.apress.com/rights-permissions.

Apress titles may be purchased in bulk for academic, corporate, or promotional use. eBook versions and licenses are also available for most titles. For more information, reference our Print and eBook Bulk Sales web page at http://www.apress.com/bulk-sales.

Any source code or other supplementary material referenced by the author in this book is available to readers on GitHub via the book's product page, located at www.apress.com/9781484243237. For more detailed information, please visit http://www.apress.com/source-code.

Printed on acid-free paper

Because the next decade expects us to know different things than the last one.

Contents

About the Author. vii

About the Contributors. ix

Acknowledgments xi

Introduction xiii

Chapter 1:   The Internet. 1

Chapter 2:   Hosting and the Cloud 7

Chapter 3:   The Back End: Programming Languages. 15

Chapter 4:   The Front End: Presentation 23

Chapter 5:   Databases: The Model 31

Chapter 6:   Leveraging Existing Code: APIs, Libraries, and
              Open-Source Projects. 45

Chapter 7:   Software Development: Working in Teams 59

Chapter 8:   Software Development: The Process 65

Chapter 9:   Software Development: Debugging and Testing 71

Chapter 10:   Attracting and Understanding Your Users 77

Chapter 11:   Performance and Scalability. 87

Chapter 12:   Security. 93

Chapter 13:   Mobile Basics 109

Chapter 14:   The Internet of Things. 119

Chapter 15:   Artificial Intelligence 129

Chapter 16:   The Blockchain. 139

Chapter 17:   Virtual and Augmented Reality 157

Index 175

About the Author

Vinay Trivedi works in technology investing with experience from Blackstone Private Equity, SoftBank Vision Fund, and Romulus Capital. He angel invests out of an ESG-oriented pre-seed fund that he co-founded called Freeland Group. His experience also includes product management at Citymapper and Locu, two venture-backed startups in London and Boston, respectively.

Trivedi serves on the Steering Committee of Startup in Residence (STiR), a program spun out of the San Francisco Mayor's Office of Civic Innovation that connects startups with government agencies to develop technology products that address civic challenges. He has also worked with the New York City Mayor's Office of the CTO on its NYCx Moonshot Challenge Initiative.

Trivedi studied at Stanford Graduate School of Business (MBA) and Harvard University (AB Honors in Computer Science), where he graduated Phi Beta Kappa and as a John Harvard Scholar, Weissman Scholar, and Detur Book Prize winner.

About the Contributors

Ashi Agrawal is a student at Stanford University studying Computer Science and spent time at the University of Oxford focused on privacy and cybersecurity. She has advised social entrepreneurs in her time with Aunt Flow and the Westly Foundation. Most recently, Ashi was a Kleiner Perkins Fellow tackling healthcare reform at Nuna Health.

Cheenar Banerjee graduated from Stanford University (BS and MS in Computer Science) where she was a section leader and course assistant for various computer science classes and a member of the Women in Computer Science core leadership team. She has worked as a software engineering intern at Facebook, Google, and Verily Life Sciences. Cheenar is passionate about furthering diversity and inclusion in the tech world.

Vojta Drmota is a student at Harvard University studying Social Studies and Computer Science. He has been involved in the blockchain industry since 2014 building applications, designing protocols, and leading educational seminars. Vojta is primarily interested in bridging the divide between the complexity of blockchain applications and their mainstream adoption.

Khoi Le is a student at Stanford University studying Immersive Design and Engineering Applications tackling the question "why is technology flat?" He has built virtual and augmented reality research, enterprise software, and entertainment experiences over the past three years. He loves singing, dancing, and playing video games.

Jay Harshadbhai Patel graduated from Stanford University (BS and MS in Computer Science) where he specialized in Artificial Intelligence and Human Computer Interaction. He works at Synapse Technology Corporation building computer vision systems for threat detection at airports and schools. Previously, he worked on Google Knowledge Graph and helped build Kite, an AI-powered programming assistant. Jay did research in crowdsourcing systems and deep learning for satellite imagery at Stanford. He is interested in making AI more understandable to ensure fair and ethical use.

Ashley Taylor graduated from Stanford University (BS in Computer Science). She is pursuing a Master's Degree in Management Science and Engineering while exploring her passion for teaching by acting as the primary instructor for several introductory computer science courses.

Acknowledgments

An investment in knowledge pays the best interest.

—Ben Franklin

This book would not exist without the lessons I learned from my parents and grandparents about the value of education and community. They raised me with the belief that curiosity and knowledge are the keys to a fulfilling life, and that we must do what we can to give back to others because we are products of our environments. These values catalyzed my interest in publishing *How to Speak Tech* and engaging deeply in the mission behind it. The book is motivated by the desire to help broaden access to STEM education to ensure that people who are "nontechnical" have the knowledge they need to be effective in a world where a basic understanding of technology almost seems required. It speaks to endeavors to retrain those displaced by AI and automation, programs that evangelize STEM education for the next generation, and organizations that battle for fair and diverse representation in STEM fields. We are committed to making a difference, and the dedication to produce *How to Speak Tech* is in service of these efforts.

I would like to express my immense gratitude to my stellar contributors who joined the team and worked tirelessly to share their expertise because they too believe in our mission. To Ashi, Cheenar, Vojta, Khoi, and Jay—a sincere thank you for the hard work you put into the chapters you contributed. They are showcases of your knowledge, skill, and passion. Thank you for letting me learn from you, and I am excited to see where your talent takes you. Ashley, thank you for taking a lead on the edits to the first edition. Your excitement for computer science education is evident and inspiring. Ashi, Cheenar, and Jay, thank you for your editing support as well, particularly Ashi who helped a tremendous amount at the 11th hour. And one special thank you from our whole team to the countless number of readers who helped us refine our chapters over the last year.

Last but not least, I am grateful to my family, friends, and mentors for their guidance over the years. Everyone's endless support and unconditional love are the real bibliography of this book.

Introduction

On January 5, 2012, Michael Bloomberg, then the Mayor of New York City, tweeted: "My New Year's resolution is to learn to code with Codecademy in 2012! Join me. http://codeyear.com/#codeyear." Codecademy, a startup that offers coding lessons to millions of users, is doing a remarkable job of making useful knowledge available for free. But why is learning how to code, of all things, among Bloomberg's top personal priorities?

Mr. Bloomberg hardly needs a job, but his tweet reflects his opinion of what the new credentials for informed citizenship and economic participation are in a society transformed by the Internet and technology. These ubiquitous forces have revolutionized how we conduct business, communicate with each other, organize our lives, order our food, plan our weddings, file our taxes, purchase our wardrobes, and much more. Technology is no longer important only to tech companies, it is shaping and disrupting every industry. No longer do only technologists require tech knowledge, everybody does. It is so important that I believe computers, the Internet, and coding should be incorporated into mandatory instruction in schools alongside reading and math.

It is neither practical nor logical that everyone learn to code sophisticated applications. But people who work in, around, or with tech—almost everybody—should understand, at least at a high level, how the Internet and its applications work. This knowledge will unlock their ability to contribute to conversations that previously may have seemed unfamiliar or daunting.

The purpose of this book, therefore, is not to teach you how to code, but rather to outline the major concepts and technologies related to each phase of a web and mobile application's life cycle. It also provides an introductory primer on each of the latest tech trends, including blockchain, artificial intelligence, and augmented and virtual reality. Whether you are an aspiring Internet tech executive, a judicious venture capital investor, or a newcomer to a company managing technology, this content is probably relevant for you. Even if you have no intention of launching an Internet company yourself, anyone living in a world in which "google" and "friend" are dictionary verbs should read this book. To too many people, the Internet and its apps are a black box, but there is no reason anyone should think that searching for a friend on Facebook is magical. As long as you understand a few key concepts, you can plumb the mystery.

As in any line of specialized work, "techies" deploy technical vocabularies largely incomprehensible to non-techie colleagues and outsiders. As a result, the gap between techies and non-techies only worsens. This dynamic is inefficient in an age of constant information overflow. If techies made a more conscious effort to demystify their jargon and if employers encouraged their non-techie employees to become more digitally literate, we could have a more informed population, greater diversity, better managers, more efficient processes, smarter investors, and more discoveries. That's exciting!

The book can be considered in two parts. The first thirteen chapters cover what I believe everyone should know about the workings of an Internet application. From the back end and the front end to promotion and cybersecurity, I see the list of topics included in this book as topics that you are likely to come across in the tech world and are always likely to be true of Internet applications. Though technology evolves rapidly, applications will usually have the same types of components with similar kinds of challenges. The tools we use might change over time, but this book focuses on the basic concepts, not on specific technologies.

The last four chapters of the book respond to the many important tech concepts that have left technical and nontechnical people alike asking questions about their applications and potential impact. This part provides introductory primers on what are some of the top tech trends to know.

The material is presented in a way for anyone to digest. With a mix of stories, analogies, and de-jargonized explanations, the book strives for accessibility. Though the material comes from academic papers, news articles, textbooks, and interviews, the tone of the book is one that should feel like a friend explaining these concepts to you in a casual, albeit slightly nerdy, conversation. There is always a tension between simplification and thoroughness; the book tries to preserve the integrity of the information while shaving away nonessentials.

The book is connected through a narrative starring you as the main character. You are building an application called MyAppoly. This application can exist in pretty much any context. If you are an entrepreneur, MyAppoly is your ticket to a $1 billion exit (hello Instagram). If you are a nonprofit executive, MyAppoly helps you fundraise from your donors. If you work at a Fortune 500 company, MyAppoly can help you stay modern and relevant. MyAppoly is anything you want it to be.

Reading this book will help you meet the new year's resolution urged by Mr. Bloomberg. Everybody knows that our world is becoming technical quickly. Every discovery is challenging us all to learn new things and question what we knew to be true and what we thought was the limit. You may not have the time to learn how to code, but how would an understanding of the basics help you? Maybe it will make you a better manager if you can relate to your developers. Maybe you will be a better investor who can differentiate between the realistic and the impossible. Maybe you will land your dream job at that tech company you recently discovered. At the very least, you should be able to understand tech news without having to consult an expert. It is time for technology to be democratized. It is time for you to learn how to speak tech.

The Internet

Most of the people in the world are connected through a global network called the Internet. Its precursor was the Advanced Research Projects Agency Network (ARPANET), which was funded by the US Department of Defense to enable computers at universities and research laboratories to share information. ARPANET was first deployed in 1969 and fully operationalized by 1975. ARPANET's solutions to the difficulties inherent in sharing information between computers formed the backbone of the global Internet.

How did ARPANET do it? Consider the following analogy. Local towns tend to be well-connected by roads. Major roads connect towns, and even larger highways connect states. To be a part of the larger national network of roads, a town simply needed to develop a single road linking the town to the highway. This road would connect the town's residents to the rest of the country. Similarly, using communication lines, satellites, and radio transmission for communication, ARPANET connected all the *local area networks* (LANs) (the town roads) and *wide area networks* (WANs) (the interstate system) together using the WAN *backbone* (the central highway).

Conceptually, we see the Internet is a network of networks—but how are these networks connected? To create a primitive local network, one can connect a few computers together with a wire, a telephone system, or a fiber-optic cable (which uses light instead of electricity). To connect multiple local networks together, one can use a "connector" computer called a *router*, which expands the size of the subnetwork. If we connect each of these local subnetworks to a single central cable, all of our local networks are connected, thus creating the Internet.

© Vinay Trivedi 2019
V. Trivedi, *How to Speak Tech*, https://doi.org/10.1007/978-1-4842-4324-4_1

Given that the Internet is essentially a physical connection of computer clusters, how does it actually function as a smoothly performing single-communication-line network? The answer to this question is where the real technical innovation of the Internet lies.

Packet Switching, TCP, and IP

Keep in mind that the whole reason researchers began investigating networks is that they needed a way to share information across computers. In the same way that highways have exit signs and speed limits, there needed to be rules to guide information flow on the Internet. To solve this problem, the Advanced Research Projects Agency (ARPA) defined how information would travel with two pieces of software: the *Transmission Control Protocol* (TCP) and the Internet Protocol (IP). These two components of the Internet's software layer operate on top of the hardware layer, the cables and devices that physically comprise the Internet.

Historically, when two computers were connected with a cable, communication was one-way and exclusive. If computer A sent something to computer B, data moved along the wire; if any other attempt at communication were made, the data would collide and no one would get what they wanted. To create a network allowing millions of people to interact simultaneously and instantaneously, the technology of the time would have required a prodigious mess of cables. The only way to reduce the need for wires was to develop a method for data to travel from multiple sources to multiple destinations on the same communication line.

This was accomplished through *packet switching*. Researchers discovered that they could break any information (such as a text, music, or image file) into small "packets"; doing so enables a single wire to carry information from multiple files at once. Two challenges seem obvious: how do all the packets make it to the right destination, and how does the recipient put them back together?

Taking a step back, every page on the Internet is a document that somebody wrote that lives on a computer somewhere. When you access a web page with your browser, you are essentially asking for permission to look at some document sitting on some computer. For you to view it, it must be sent to you—and TCP/IP helps do just that.

TCP breaks the document into little packets and assigns each one a few tags. First, TCP numbers the packets so they can be reassembled in the right order. Next, it gives each packet something called a *checksum*, which is used to assess whether the arriving data were altered in any way. Last, the packet is given its destination and origin addresses so it can be appropriately routed. Google Maps can't help you here, but IP can. IP defines a unique address for every device on the Internet.

So these packets have your computer's IP address. IP helps route these packages to their destination and does so in a process much like the US Postal Service's method for delivering mail. The packets are routed from their start to the next closest router in the direction of the destination. At each step, they are sorted and pushed off to the next closest router. This process continues until they arrive at their destination. In this way, IP software allows an interconnected set of networks and routers to function as a single network.

One prized characteristic of IP is the stability guaranteed by network redundancy. If a particular segment of the overall network goes down, packets can be redirected through another router. When the packets reach you, TCP verifies that all packets have arrived before reassembling them for viewing.

Several computer companies had actually already developed independent solutions to the problem of computer-to-computer communication, but they charged people to use them and they weren't mutually compatible. Why buy company A's solution if it did not connect you to people who bought company B's solution? What distinguished ARPANET was that ARPA published its results for free, making TCP/IP publicly accessible. When the US military branches adopted TCP/IP in 1982, ARPA's open and free approach was legitimized and institutionalized. Ever since, computers could only send or receive information via the Internet if they had TCP/IP software installed.

Thus, starting in 1982, researchers could share information around the world, but the question remained how to display and view that data. In 1990, Tim Berners-Lee and others at the European Organization for Nuclear Research (CERN) developed the precursors of *Hypertext Markup Language* (HTML) and *Hypertext Transfer Protocol* (HTTP), which jointly enable the exchange of formatted and visually appealing information online. After 1991, the first graphical browsers—programs that allow you to view web pages on the Internet—were released. This created an attractive and efficient way to share and consume information on the Web.

Around this time, the cost of personal computers dropped, and online service providers such as AOL emerged, offering cheap access to the Internet. This confluence of factors led to the Internet's rapid growth into the network we use today.

HTTP and Using the Internet

How data travel physically is pretty straightforward and defined by the protocols developed by ARPA—but how do you tell someone to send the data in the first place? What is actually happening behind the scenes when somebody—let's say you—visits your brainchild, MyAppoly?

> ▧ **MyAppoly** In case you skipped the preface, you should be aware that this book is structured as a loose narrative starring *you* as the main character. The premise is that you are building a web application called MyAppoly. The name is just a catchall; I encourage you to imagine MyAppoly in any context that catches your fancy. If you are a killer app entrepreneur or angel investor, MyAppoly will be your ticket to a $1 billion exit strategy. If you are a nonprofit executive, MyAppoly will help you raise funds and connect your volunteers. If you work at a Fortune 500 firm, MyAppoly will help your company stay competitive and ahead of the curve of evolving consumer expectations.

You open your web browser to access a picture on the MyAppoly website. In doing so, your computer becomes the *client,* requesting information. The physical web pages you visit are documents usually encoded in a language called HTML stored somewhere on a computer called the *server.* All of the files of your application, including pictures and videos, live on the server. These files are referred to as *resources.* Because a *client,* the user, is accessing a *server,* the Internet is said to follow a *client-server architecture.*

You proceed to type your web address—the *uniform resource locator* (URL), www.MyAppoly.com—into the browser. Technically, you could have typed the specific IP address of the MyAppoly server, but who has the capacity to remember the IP address linked to every website? The *Domain Name System* (DNS) converts human-friendly domain names such as MyAppoly.com to the computer-friendly IP addresses.

You reach MyAppoly.com and click a link to view the picture gallery on the website. Remember, all of these pictures also live on the server. Let's say that they are all in a folder called "Pictures." If you click the first picture, you are taken to http://www.MyAppoly.com/Pictures/pic1.jpg. The URL's component parts indicate that we're using the HTTP protocol, the proper server (through the domain name), and where on the server the files are located (in tech-speak: the *hierarchical location* of the file). In other words, the URL is the text address of a resource.

How exactly do you receive these web pages? First, the client—your browser—accesses the DNS to obtain the IP address corresponding to MyAppoly so it knows where the server is. Your browser does not physically travel to the MyAppoly server to fetch the picture, so it must send a message over the Internet telling the server to send the file. An HTTP request does just that. HTTP is a set of rules for exchanging resources (text, images, audio, and so on) on the Internet.

Your request can be one of several different methods, most commonly GET or POST. The GET method tells the server that the client wants to retrieve files from the server. On receiving a GET request, the server retrieves the appropriate files and sends them back to your browser. The other request type is POST, which the browser uses if you are sending data to the server. In some instances, either method could service the request, but they are different in how data are actually sent over the Internet. With the GET method, the information you send to the server is added to the URL. If you are searching for the word "mediterranean" on MyAppoly, for example, the GET request redirects you to the URL www.MyAppoly.com/search?q=mediterranean. If the search term is sent via POST, the term would be within the HTTP message and not visible in the URL. It is considered good practice to reserve POST requests for those that alter something on the server-side.

So the client has issued a request, which finds the MyAppoly server and tells it to GET the page containing the "pic1" file. The server fetches the resource and sends it back as a *response* using TCP/IP. The browser can use the header's information to display, or *render*, the resource. The process is not necessarily finished, however, because the client may need to send more requests. Since the server can only send one resource at a time back to the browser, several requests may be needed to retrieve all the resources required to construct a web page. If you want to view the page that has the pic1 picture on it, you are asking for two resources: the HTML page, which has the text content and layout of the page, and the pic1 image. Therefore, the browser needs to send at least two requests.

Conclusion

With your knowledge of some of the basic elements, operations, and tools of the Internet, you are probably eager to move on to the challenge of creating MyAppoly. However, you must first set up website hosting, which is required for your application to be on the Internet.

Hosting and the Cloud

Availability for a brick-and-mortar retailer includes inventory and the physical placement of products on shelves. In some ways, the Internet equivalent is *hosting*, the process by which the creator makes a website accessible to users on the Internet.

It's almost entirely impossible to escape a conversation about the Internet without mentioning the word *cloud*. But what does "the cloud" refer to, and why has it become a ubiquitous term?

Hosting

As you learned in Chapter 1, when you visit a website, your browser sends a message to a server asking for a file. How did the server have the information in the first place? The process by which a website is put on a server and made available on the Internet is called *hosting*. Your website files are saved onto servers that are connected to the Internet nonstop, ensuring the website is always accessible. Imagine if you put your files on a server that someone turned off every night. Nighttime visitors and customers would receive error messages, meaning lost business—so an always-on high-speed Internet connection for the hosting server is a must. In addition to internet connection, servers require special web hosting software but I don't get into that here.

© Vinay Trivedi 2019

V. Trivedi, *How to Speak Tech*, https://doi.org/10.1007/978-1-4842-4324-4_2

You probably don't want to host a website yourself. You certainly *could*—it would be free (electricity and Internet not included), and you would have complete flexibility over the server. The downside is that you have to manage it, which requires significant technical knowledge.

Enough people in the world have realized that self-hosting is troublesome, so smart entrepreneurs started companies that provide hosting. They offer computers designed to store websites optimized for accessibility and speed. They removed all unneeded components on computers so they can be more efficient and cheaper servers. Say goodbye to card games and paint programs. No more monitor or keyboard either. Hosting providers accumulated racks of these minimalist computers and rented space on these servers to people who wanted to host their websites.

Hosting providers quickly became popular because they allowed website developers to focus on their product. Hosting providers typically offer a control panel for website developers to manage the site and allow files to be moved to these remote servers using *File Transfer Protocol* (FTP), special software that allows you to upload your website files to the servers.

Hosting Considerations

When choosing a hosting provider, consider the following parameters:

- *Server type*: Different servers have different web hosting software (such as Apache, OS X Server, and Windows Server), which does not matter for the average user and use case.

- *Disk space allowance*: The size of your website and database will determine how much memory you need on the server.

- *Bandwidth*: Every time someone visits your web page, the browser requests files from the server. The amount of data transferred from your server back to the user is known as *bandwidth*. The more users visiting your site or the more images and resources your web page has, the higher the bandwidth. Some hosting providers base a portion of pricing on bandwidth needs since it costs more to serve 1 million customers than just one.

- *Reliability and uptime*: Make sure the hosting provider keeps your website on the Internet and accessible as close to continuously (24 hours a day, 7 days a week, 365 days a year) as possible.

- *Price*: Shop around for different options. The prices vary among the categories outlined next, but within a particular category, prices may vary just a little. Have a realistic budget in mind that balances your requirements with available packages.

The Different Types of Hosting

Hosting providers make their servers available for use in many ways. Below are a few:

- *Shared hosting*: When multiple users share a single server, it's called *shared hosting*. The hosting provider essentially divides a server into several parts and rents each part out to a different user. Because the server is shared, this type of plan is typically the cheapest but also the most inflexible. You can't update software or change configurations on the shared server, as your changes might affect the other websites running on the same server (and their changes would affect you!). Despite the trade-off in flexibility, this cheaper solution typically works for simple websites, such as personal websites.

- *Dedicated server*: As the name suggests, a dedicated server plan allows you to rent an entire server to do whatever you please. You can customize the server in any way you want, and you can ensure that other people's problems do not become yours, but a substantial increase in price accompanies the added flexibility and control. This type of plan makes sense if you are hosting several websites, especially if they have significant traffic and substantial database use (usernames, credit card information, popular items, etc.).

- *Virtual dedicated server*: A virtual dedicated server, aka virtual private server, is in some ways a combination of a shared hosting plan and a dedicated server plan. Hosting providers can offer server space that you can treat as a dedicated server, but the server is really shared. You can customize it however you want, unlike in shared hosting plans, yet several customers can use the same server. Computer space and processing power are not wasted, thus benefiting the hosting provider, and customers do not have to pay for an entire server if they don't need it.

- *Collocated hosting*: In a collocated hosting plan, you own the server, but delegate its management to a hosting provider. What you pay for is bandwidth (Internet usage) and maintenance fees (for cooling and so on).

The Cloud

The cloud is best understood when you realize its similarities to the mainframe computers of the 1950s. These computers were massive, pricey, and available only to the governments and large organizations that could afford them. Rather than work directly on the computer, users would access the mainframe using a dummy computer with a special, text-only program called the *terminal*. Through these dummy computers, users would gain access to the real computer, where all the data and programs lived. All information and functionality were centralized, in a sense, but as computers become cheaper, the idea of a personal computer began to take over.

However, individual computers are limited by their memory and data processing speed. As network technology improved, computer scientists realized that connecting computers in a network would dramatically improve their capabilities. This network of computers is called the cloud. The decentralized computers that comprise the cloud can interact collaboratively through the Internet. In some ways, the cloud is a reversion to the mainframe computers of the 1950s—users interact with the fast and powerful cloud through an interface on their individual, much less powerful computers.

When someone says, "I stored it in the cloud," or "I accessed it in the cloud," they're referencing a powerful tool. Through the Internet, they accessed some service (such as Dropbox) and stored their files there. Rather than saving it on their hard drive or relying on offline services, the cloud leverages the Internet to make resources and services available. In addition to photos, music, and any other type of information you can imagine, the network of servers that forms the cloud also stores everything from simple websites to more complicated web software that can be accessed online.

Cloud computing refers to the access and use of information and software on the cloud by computers and other digital devices such as smartphones and tablets. It encompasses the idea that you no longer have to save everything to your physical computer. Cloud computing means that you don't need to go to a store to buy software on a disk that you have to insert and install onto your machine. Today, most companies host some version of their product online.

According to the US National Institute of Standards and Technology (NIST), the cloud computing model "is composed of five essential characteristics, three service models, and four deployment models."[1]

---

[1]Peter Mell and Timothy Gance, "The NIST Definition of Cloud Computing." National Institute of Standards and Technology Special Publication 800-145, 2011. Available at http://csrc.nist.gov/publications/nistpubs/800-145/SP800-145.pdf.

The five essential characteristics of cloud computing identified by NIST are the following:

- *On-demand self-service*: You, the client, should be able to get access to the cloud's resources and software whenever you want. To log on to Salesforce.com, all you need is a computer with Internet connection and a valid username and password.

- *Broad network access*: The cloud should be accessible across a broad range of client platforms (whatever devices you are using to access the resources and/or software—tablet, smartphone, laptop, etc.).

- *Resource pooling*: Computers in the cloud work with each other to support multiple users and solve problems too large for isolated computers.

- *Rapid elasticity*: Because computers in the cloud can talk to each other over the network, if specific computers are overwhelmed with traffic, other computers can pick up the slack. The cloud allows capabilities to be scaled up and down rapidly.

- *Measured service*: In the traditional software model, customers pay a set amount of money to install and to use a program. Whether you used it every day or wanted to try it out, it would cost you the same. Now, companies can measure a customer's usage in terms of storage, processing, bandwidth, and other metrics. Pricing based on usage is now possible, which leads to a more efficient market.

The three service models for cloud computing identified by NIST are the following:

- *Software as a Service (SaaS)*: Any application that a user can access through a browser, such as Dropbox, falls under SaaS. Also related are APIs, which are discussed in more detail in Chapter 6. Because the service is offered over the cloud and a copy of the application is not installed on every user's computer, we can say the cloud is based on a multitenant architecture.

- *Platform as a Service (PaaS)*: This term will be familiar to people who have been exposed to cloud platforms for the programming community, such as Heroku. These platforms provide environments for users to build their own software. Among other things, they provide web servers for you to build your own web applications.

- *Infrastructure as a Service (IaaS)*: The growth of the cloud has created an entire business out of cloud infrastructure, which allows users to host their websites on the cloud. This enables   lot of efficiencies. As an example, after launching

  After launching a new promotion, you might expect a lot of traffic. Historically, you would need to purchase enough servers to serve your site during this peak time. After the surge of visitors, you would be left with a lot of expensive servers and relatively low traffic. The cloud transforms these fixed costs into variable costs. Because you can add and take away server capacity easily with the cloud, you pay for only what you use.

  Amazon, which has built up data centers to power its own website, made its additional servers and tools available to others through a service called Amazon Web Services (AWS). Amazon Elastic Compute Cloud (EC2), a part of AWS, allows you to remotely and flexibly add or remove servers according to your plan. Microsoft, Google, and many others offer similar services.

NIST also identifies the four deployment models for identifying who can have access to the information in the cloud, which are outside the scope of this book.

Benefits of Cloud Computing

The benefits of cloud computing include the following:

- *Accessibility and independence*: You can access the application or data on any device connected to the cloud, so you are no longer tied to your physical computer to get what you need.

- *Backup*: Because your files are stored securely in the cloud, if you drop your laptop in the bathtub or lose your tablet in the subway, the data are safe and sound.

- *Prorated pricing*: Pricing is metered or based on usage.

- *Lower hardware costs*: A lot of the computation is taken care of by the computers in the cloud, so you do not necessarily need the fastest computer with the most up-to-date hardware. All you need is a computer with reasonable Internet speed. You also don't need to store all the information yourself, because your files can all live on the cloud. Inexpensive laptops such as Chromebooks only provide fast Internet access.

- *Better performance*: Because the cloud reduces the need for you to install feature-redundant programs, your computer has more free memory to work faster at other things.

- *Fewer maintenance issues*: Because software lives on the cloud, you are a lot less likely to have problems of installed software conflicting with other software or newer versions of the same software.

- *Access to the latest version*: When you visit a website, the version of the application you access will always be the most up-to-date version.

- *Easier group collaboration*: With everyone on the cloud, groups can collaborate more effectively. Think Google Drive and document collaboration.

- *Increased computing power*: The computation power of a single computer is limited. On the cloud, however, processing power can be spread across the entire network.

Disadvantages of the Cloud

If you are on cloud nine after reading the long list of cloud benefits, I am sorry to have to rain on your parade. The cloud is not perfect for the following reasons:

- *Internet dependency*: If the Internet is down or you don't have a reliable Internet connection, you cannot access the data and software you might need. Many cloud applications allow you to download your data as a work-around, although some of them charge you to do so.

- *Security and privacy vulnerability*: The cloud is becoming more secure, but companies are justifiably wary about having their data accessible via the Internet. If your data exist with someone else, are you keeping your information private? Despite increasingly sophisticated security defenses, vulnerability to malicious attacks and hacking thefts is a growing worry at all levels (Chapter 12).

Conclusion

The Internet proliferated in large part because hosting became so easy. Entrepreneurs made their creations available to their customers quickly and inexpensively by utilizing the latest hosting technologies. As consumers have started to shift their preferences toward cloud services, newer tech companies are often finding themselves in an attractive light versus large companies who may be too tied to legacy systems to take full competitive advantage of cloud benefits.

It goes without saying that you need to build a website before you can host it on the cloud. Whether you are an opportunistic entrepreneur aiming to modernize old software or you work at a large company hoping to stay current with new web technology services, you must understand the parts of a web application. You will acquire this understanding by building MyAppoly, as you will start to do in the next chapter.

The Back End: Programming Languages

You have spent the last couple of weeks brainstorming and refining the vision for MyAppoly. What's next? Many would-be products never make it past this question due to the technological barrier of creating an application. Software applications are written or *coded* in a programming language. The code defines how the application will run and respond, encompassing everything from what you see to what happens in the background. One of the first technical decisions to make is to choose a programming language for the product. You may have heard of languages like Python, Ruby, and C, but what are they, and what are the significant differences between them? What factors should you consider when selecting a language?

A programming language is no more than a set of commands that a computer can follow. Programming is the art of combining these commands into a set of instructions that define the application's behavior. The MyAppoly team will first need to use a programming language to define the *back end*, which is responsible for processing all of a user's actions. For example, when you click "Add Friend" on Facebook, its back end processes the click by interacting with a database and other sources of information to deliver what you are expecting. In this example, the back end sends a message to your friend. For

Google, the back end might take your search query terms and pass them through its algorithms to find relevant results. The back end is also called "server side" because the code defining the back end lives on a server. The user interacts with the back end through the *front end*, the interface and range of interactions you can perform on a given page. The "Add Friend" button you clicked and your friend's profile page are part of the front end of Facebook. It's the part you see. We discuss the front end in more detail in Chapter 4.

Programming languages are created and developed at an astonishing rate. For this reason, an understanding of the basic technical terminologies and classifications of programming languages will help you keep up with the speed of development. The language you ultimately pick will be based on your needs and your particular situation.

The popularity rankings of programming languages fluctuate from year to year. For example, the top ten programming languages in August 2018 show the following ranking changes compared to August 2017:[1]

1.	Java	(same)
2.	C	(same)
3.	C++	(same)
4.	Python	(up from #5)
5.	Visual Basic .NET	(up from #6)
6.	C#	(down from #4)
7.	PHP	(same)
8.	JavaScript	(same)
9.	SQL	(previously unranked)
10.	Assembly language	(up from #14)

Classifying Programming Languages by Level

Computers only understand binary, a system of ones and zeros. All files, programs, and data are fundamentally made up of ones and zeros. Each one or zero is called a *bit*, and a group of eight bits is a *byte*. Computers interpret bytes depending on how a programmer chooses to label or specify them. For example, let's consider the byte 0100 0001. As a number, that byte represents

---

[1]TIOBE Software BV, www.tiobe.com/index.php/content/paperinfo/tpci/index. html. Note that not all of these languages relate to the back end.

65, but you could instead tell the computer to read it as a character, and it would be interpreted as an A. Larger pieces of information, such as the online version of this book, are comprised of a huge number of bytes together (measured in kilobytes, megabytes, or gigabytes). For example, when purchasing a computer, you might see that it has 16 GB of space. This means the computer can store 16 gigabytes of information.

Programmers can manipulate bits and bytes through a computer's machine language, but imagine coding MyAppoly by changing ones and zeros. It would be extremely difficult, tedious, and costly. Thus, although machine code is closest to what the computer actually understands, programmers defined simple functions to do the bit coding for them. Together, these functions form *assembly language*. A programmer can code in assembly language and use an assembler to convert the program into binary for the computer.

Assembly language proved to be tedious as well, so programmers created *high-level languages* with more powerful commands. These languages allow us to write code that more closely resembles human language. High-level language code is converted to machine language directly or by way of assembly language. As a result, higher-level languages don't expand the set of computer behaviors possible to express in code. The difference between using a higher-level language and assembly is similar to the difference of listing only a few instructions for making a sandwich and tediously explaining the process in a hundred small steps.

High-level languages, such as PHP and Python, are called *high level* because they enable the programmer to think abstractly rather than in specific bits and bytes. It follows that assembly languages are considered *low level* because they are closer to the hardware. Some languages, such as C, are considered *middle level* because they allow a programmer to think about bits but relate more closely to human language and have more complex commands than do assembly languages.

Processing High-Level Languages

For coding MyAppoly, you will use a high-level programming language. High-level languages may be subdivided by various criteria. One useful taxonomy, based on how a language is read by computers, divides high-level languages into two types, *compiled* and *interpreted*.[2]

---

[2]Many experts identify a *hybrid* of compiled and interpreted as a distinct third type of implementation. See, for example, Robert W. Sebesta, *Concepts of Programming Languages*, 8th ed. Addison Wesley, 2008.

Compiled languages use a compiler to translate the code into a language that the computer's processor understands. If you code MyAppoly in C or Java, you are using a compiled language, and the program you code is called the *source language*. The compiler essentially translates the source language into machine-executable code, much as one might translate an English book into French. Compiled languages are fast to run but slow to develop, i.e. write an application in (Chapter 11.)

Interpreted languages such as PHP are translated in real time. They are interpreted much as an English speaker might have a live interpreter translate continuously and simultaneously to someone who only speaks French. As a result, the execution is one to two orders of magnitude slower than in compiled systems. Compare the time it would take a French speaker to sight-read a French translation of this book with the time it would take the same French speaker to listen to this book being read aloud by a French interpreter translating live from the English original. A less obvious disadvantage is that interpreted languages are less efficient because they often use more memory. Although they are slower to run, interpreted languages tend to be much quicker to develop.

Other Taxonomies of High-Level Languages

Coding experience makes it easier to understand the following descriptions, but the main takeaway is that different languages are built for different types of tasks. One language may be suitable for building a complicated trading algorithm for a financial investor, whereas another language may be best for building a simple tool for a teacher to sort through a list of students for grading.

Codifying the different types of programming languages is not simple. If I asked you to classify all the types of dogs in the world with a single system, you would find it a difficult task. There are different ways you could try, but inevitably there would be overlap between the various groups. As with Labradoodles and Cockapoos, the overlap between languages is extensive. Nonetheless, the following taxonomies of high-level languages contain keywords you will most often hear in the tech world:

- *Imperative languages* focus on *how* to achieve an objective and are typically written as a sequence of commands that create and modify variables. Examples include C and Java. These languages fall in contrast to *declarative languages* that focus on the *what* of an objective and leave the *how* to the programming language. Examples include SQL and regular expressions.

- *Functional languages* rely on functions and recursion as their main method of computing instead of stored variables or iteration. Examples include CAML, Standard ML, Scheme, Lisp, and Haskell.

- *Object-oriented languages* create *objects* to store information and execute commands. For example, in an academic application, a programmer might create an object for each student. The object contains *variables* that define data like the student's name and advisor and a set of related functions, called *methods,* which describe how the computer should retrieve the student's grades or display their courses. Everything that might usefully describe the student is defined in this student object. Object-oriented programming also implements *abstraction,* which allows a programmer to define shared attributes in one place. Suppose you now wanted to create a teacher object. It would include a variable to store the teacher's name and a way to display their courses. Note that these are the same things we defined in our student object. Rather than rewrite your student variables and methods for teachers, you can create an "academic person" object and have students and teachers be specific types of "academic person" objects. Examples of object-oriented languages include Java and C++.

- *Scripting languages* share many qualities with imperative and object-oriented programming. These languages are weakly typed, a term that references the flexibility with which they define and convert between types. Recall that variables need to be specified as a particular type, such as a number or character, so the computer interprets its underlying bytes the way we want. This flexibility allows for rapid development but reduces reliability (discussed in the next section). Scripting languages such as Python and Ruby interface easily with many different systems, such as APIs and databases (Chapter 6).

- *Parallel programming languages* allow for multiple execution sequences to proceed concurrently. Typically, parallelism is used to speed up a process—evaluating two things at once is quicker than doing one after the other—or to react to events that might occur at the same time.

- *Query languages* are used to interact with a database (Chapter 5).

- *Markup languages* use *tags* and special characters to give structure to a document (Chapter 4). This structure is typically used to visualize a document. Examples include HTML and XML.

Choosing the Right Language

You now have a better understanding of programming languages and their different types, but you may not be closer to picking a language for MyAppoly. Language design principles serve as useful background in evaluating language effectiveness, but practical factors will ultimately govern your decision.

Technical and Design Considerations

In evaluating programming languages for their suitability to your requirements, first consider their technical and design parameters, as follows:

- *Naturalness of application* refers to the relative ease of coding your application in a particular language based on the functionalities and constructs it provides. Object-oriented programming languages became popular in part because they allow the programmer to think about the solution to the problem in many ways and therefore fit many programmers' uses.

- *Reliability* refers to a program's ability to perform the same way each time. A program can run into different types of errors (relating to how a computer interprets the bytes that make up a variable) either when you run the program (at "runtime") or when compiling. It is cheaper to catch errors at compile time rather than at runtime, but as we mentioned before, compiled languages typically take longer to develop and execute. The trade-off between cost of execution and reliability is one that programmers constantly face (Chapter 11).

- *Support for abstraction* (as discussed in the previous section) refers to the ability to define and use complicated structures or operations in a way that allows many of the details to be ignored.

- *Portability* refers to the ease with which your program can be translated into another programming language. For example, if you are choosing between two new programming languages, you may decide to use the one that can more easily translate into a well-established language.

- *Efficiency* refers to the time that the code takes to execute, often at a cost to abstraction. Lower-level languages allow more control over memory usage and provide a more optimized experience for users (Chapter 11).

Practical Considerations

The following practical considerations will ultimately govern your selection of the programming language for building MyAppoly:

- *Applicability*: If you have the wrong tool for the job, your code might become unnecessarily complicated.

- *Documentation*: Documentation refers to resources that explain a language's syntax (what specifically you need to type to get it to work), usage, and constructs. Languages that have thorough documentation are useful, especially if you and your team are learning to code as you go.

- *Development time*: Historically, speed and memory usage were key factors in choosing a language, because processors were slow and memory was limited. Now, however, hardware is cheap and becoming cheaper. Because hardware is now less expensive than hiring a new programmer, organizations tend to prioritize development time over execution time.

- *Reliably updated*: More popular languages, such as PHP and Python, tend to be updated regularly based on programmer feedback. Each update makes it easier to express instructions, expands the set of commands a programmer can use, or fixes issues in the language.

- *Maintainability*: This practical factor relates to a language's simplicity and readability. If MyAppoly is to last for a meaningful period of time, you will need to find new programmers to satisfy your growing needs. If understanding, updating, and correcting your code is a nightmare, your product can be severely delayed.

- *Committed community*: Another benefit of popular languages is the community of programmers using it. If many people use the language, the chances are good that the behavior you are trying to implement has already been programmed by someone else. Forums and blogs

(Stack Overflow is a popular example) are full of sample code to help you get your product up faster and more painlessly.

- *Talent pool*: The committed community also comprises your target market for technical hires. If you choose an obscure language, your ability to scale up your team with people that know the language intimately will be limited.

- *APIs, libraries, and tools*: A large community of programmers might post tools that you can directly plug into your program to save the time of programming it yourself (Chapter 6). These tools support the development of web applications by equipping developers with functionalities that are common among all web applications.

- *Integrated development editors (IDEs)*: This one-stop application has an editor for typing code, a source compiler to translate the code, a debugger to help spot any errors in the program, and a few other project management–related tools. The *programming environment*, or the collection of tools used to develop a product, is important, and choosing a language with better support can help your programmers be more efficient.

Conclusion

While debating which language to use for MyAppoly, you realize that tech startups have increasingly begun to prioritize development time over almost all other considerations. There are two primary reasons: hardware is constantly becoming cheaper (Chapter 14) and popular product development doctrines dictate rapid iteration (Chapter 8). Your competitors are building as fast as they can. Conscious of your need to be the first mover, you expeditiously consider all of the factors set out in this chapter and settle on the most popular scripting language of the day. You have little time to waste.

The Front End: Presentation

Knowing which language to use is helpful, but without a front end, a user cannot do anything. Recall that the front end is the interface for users to interact with the back end. It refers to what you see and the range of interactions you can perform on a given web page. You start to think about how MyAppoly will look, which pages you want, and how they will be connected (e.g. button A on page 1 takes you to page 9). Then, you finalize a set of features you want to include in MyAppoly and develop wireframes of each web page.

Wireframes are simple blueprints for a screen. Putting together wireframes forces you to think about information design and interaction design. *Information design* relates to how you present all of the content or messaging you want to communicate to the user. *Interaction design* refers to the series of actions you want the user to perform to complete some overall task. For example, the placement of product descriptions and reviews on a site such as Amazon falls under information design, while the flow through which a user reviews an item, adds it to the cart, and ultimately purchases it relates to interaction design.

As you design your front end, you'll also want to consider the user interface and the user experience. The *user interface*, frequently abbreviated as UI, refers to how the application looks. A good UI makes an application intuitive to use, whereas a bad UI may be cluttered or confuse the user. The *user experience* (UX) refers to how users feel about using your application; you should always strive for a positive user experience, which generally comes from having a

©Vinay Trivedi 2019
V. Trivedi, *How to Speak Tech*, https://doi.org/10.1007/978-1-4842-4324-4_4

back end that works well, an easy-to-understand UI, and information and interaction designs that are well thought-out.

You review the mock-ups for MyAppoly with your CTO. In effect, you have created a paper version of your website. Great! How do you get your website to look like that? How will you create the interactive effects that you see on most modern web pages? Certain languages and tools have become so ubiquitous that it is worth learning the basics because together they form the presentation layer and define what your users will experience.

Front-End Technologies

The standard front-end tools for creating the presentation layer include HTML/XHTML, CSS, JavaScript, and Ajax.

HTML

HTML is a markup language that defines the content and structure of web pages. As discussed in Chapter 3, a markup language uses tags and special characters to give structure to a document. There are two types of tags: container and standalone. Container tags flank text that you want to format or contextualize. Take the example of `<h1>Read my book</h1>`. When a browser translates the markup code in the web page that you see—a process called *rendering*—it will see the starting `<h1>` tag and know to bold and enlarge the text that comes after it. Thus "Read my book" will look like a heading. `</h1>`, the end tag, tells the browser to no longer bold any text and to continue reading the code until it hits the next tag. Such tags can also be nested, such that if you wanted your heading to be italicized, you could write `<h1><i>Italicized Heading</i></h1>`. Whereas container tags describe formatting with both start and end tags, standalone tags are those which you place into the code to create some type of effect. When a browser reads `<br>` in HTML, for example, it knows to create a line break in the web page before continuing to read the rest of the code.

Thus, markup languages use combinations of tags to define the structure of a document. The style sheet (which will be discussed with CSS) adds style to the tags so rendered documents don't look so bare. Without this styling, web pages would all look the same, with the same font, font size, and color. Style sheets tell the browser things such as "Make all those `<h1>` tags a different font!" Thus the markup document and style sheet together form the bare essentials of web pages.

With that introduction, let's jump back to HTML. In the 1980s, Tim Berners-Lee used a *metalanguage* called *Standard Generalized Markup Language* (SGML) to create HTML, the main markup language used to display web pages on the Internet. As with other markup languages, HTML uses tags that are hierarchically nested to give structure to text that forms the web page. Recall that elements can be nested, in the same way that the italicized element was contained within the <h1> tag described previously. On visiting a web page, a user is essentially opening an HTML file, which the browser then renders into a more legible format.

The structure of HTML pages typically consists of the following six parts:

- *DOCTYPE*: Over time, new versions of HTML and XHTML (discussed later) have been developed. Therefore, we have the problem where web pages are written in one of several versions and types of markup languages available. When you go to your favorite website, how does your browser know which markup language created that web page? If it did not know, it might interpret one of the tags incorrectly or might not present the page at all. Therefore, DOCTYPE, or Document Type Definition/Declaration (DTD) tells the browser in which markup language the file was written.

- *HTML*: The HTML tag contains all the tags or elements of the document (except the DOCTYPE). It defines the area that the browser should render.

- *Head*: The head element contains the information that the browser uses to read the rest of the document. In addition to the title of the document, the head element includes other scripts, meta-information, and style sheets—all discussed later.

- *Title*: This defines the title of the web page. In addition to communicating information to a user, a good title is one of the keys to *search engine optimization* (SEO), the process by which you can improve your position in search engine results (Chapter 10).

- *Meta*: Metatags are used to convey information to the browser such as the keywords of the page or a description of the page content. These tags are also important for SEO.

- *Body*: The body element contains the actual page to be read including all text, images, links, and so forth.

Numerous other types of tags exist that introduce graphical images, tables, lists, and almost anything else you can think of. Tags can have attributes that give more descriptions about the element, such as information used for styling. The newest version, HTML5, also supports new features such as animation, geolocation, touchscreen support, and content handling.

The following two sections discuss other terms you are likely to come across when discussing markup languages and the design of web pages.

XML and XHTML

XML is a markup metalanguage, which means it is used to define other markup languages, so in some ways it is similar to SGML. It is much simpler, however, than SGML, the technology from which HTML was developed. XML consists of two parts: a document and a style sheet. The document contains tags or elements, as described earlier, and the style sheet gives style to the various elements.

Why do we need XML? SGML includes a lot of complex and often unnecessary aspects. As a result, HTML is often forgiving of errors and is more lenient with rules, leading to different renderings of the same HTML document by different browsers or devices. It becomes difficult for you as a developer to control the user interface if each browser reads your work differently. XML is stricter and has guidelines that need to be followed. Therefore, if we were able to rewrite HTML with XML, we could create a markup language that might not run into as many problems as HTML. This idea inspired the markup language called XHTML. With stricter coding guidelines, XHTML pages can be more accessible on different devices and browsers, though XHTML is not *backward-compatible*, meaning old browsers may not be able to read newer versions of the markup language.

CSS

Cascading Style Sheets (CSS) is a language that tells the browser how to style and position the various HTML/XHTML elements. CSS syntax is brief. It includes a selector that tells the browser which HTML tags to style, and a declaration that tells the browser how to style the elements.

There are three places where CSS styling can live. First, it can be placed in-line as in `<h1 style="color:blue">My blue heading </h1>` tells the browser to color the heading blue. Alternatively, all style codes can be aggregated at the top of the page within the `<head>` tags. The browser is smart enough to match the style code to its corresponding HTML tags. Lastly, all CSS can be put in an external sheet that is linked to within the HTML page. You specify which HTML tags you would like to style, and the browser can manage the rest. Of the three options, the external CSS file is considered best practice for a number of reasons:

- *Readability*: It is better to keep structure and style separate.

- *Speed*: Faster load time.

- *Maintenance*: Imagine you want to update the style of your entire site. If the CSS were in-line, you would have to go through all the HTML to update all of the styles, but if everything lived in an external file, you could easily make updates. You could even create an entirely new file called "styleversion2" and then change the link in the HTML document. Depending on your mood, you could change the link from styleversion1.css to styleversion2.css and the entire feel of your website could change. This powerful capability resembles *abstraction* (Chapter 3).

Inheritance, discussed in Chapter 3, exists in CSS, whereby the style of *parent* elements is passed on to *children* elements. In other words, an element within another element inherits the style of the larger container in which it is contained. We call this idea of inheritance *cascading, hence the C in CSS*.

CSS contains eight general style categories as follows, together with examples:

- *Type*: Font, color

- *Background*: Color, image

- *Block*: Word spacing, line spacing

- *Box*: Width, height, float

- *Border*: Style, color

- *List*: Style, position

- *Positioning*: Position, height, visibility

- *Extensions*: Page break, cursor

Positioning in CSS is based on a box model whereby every element can be described as being located in a nested zone, ranging outward from *content* in the center, through *padding*, and then on to *borders* and *margins*. In some ways, designing a web page with CSS is nothing more than arranging boxes on a screen.

JavaScript

With HTML and CSS, you can create great-looking websites with content formatted nicely, but they will lack interactivity. Drop-down menus, informational boxes that appear when you hover the mouse, and so much more can be attributed to JavaScript, an object-based scripting language (remember the definitions from Chapter 3?). It lives in the HTML document and is rendered by your browser (the *client* trying to access the site); therefore we say JavaScript requires *client-side* processing. This is why JavaScript, unlike the other scripting languages we have discussed, is considered front end and not the back end. It is not executed by the server.

First introduced by Netscape, JavaScript is now ubiquitous across the Web and will serve to make MyAppoly interactive and animated. Let's say you want an informational pop-up window to be displayed after the user clicks on the MyAppoly logo. How is this accomplished? The HTML element where the logo is contained will contain some reference to a piece of JavaScript code called an *event handler*. Event handlers "wait" for certain "events"—actions that the user completes—at which point the computer starts following the corresponding JavaScript instructions. In this case, the JavaScript event handler will create a pop-up window and add it to the user's display when the logo is clicked. Thus, events, event handlers, and functions define a large part of how JavaScript can be used to add interactivity to a web page.

As with CSS, JavaScript can be placed either directly into the HTML document in the <head> element or in an external *.js* file with a link from within the HTML document.

Document Object Model and DHTML

What if you wanted to change the color of the background when the user clicked the MyAppoly logo and produced a pop-up window? This is slightly more complicated because JavaScript must not only create a pop-up window but also alter the HTML/CSS code that defines the background color of the page. This *Dynamic HTML* (DHTML) combination of static HTML and JavaScript is enabled by the *Document Object Model* (DOM). The *World Wide Web Consortium* (W3C) defines DOM as "a platform- and language-neutral interface that will allow programs and scripts to dynamically access and update the content, structure and style of documents." The DOM essentially defines all the elements on an HTML page as objects, places them in a tree-like structure based on how the elements are nested so you can easily reference them, and makes every element on the page interactive. You can theoretically reference any element in the document by tracing the path from the top of the tree to that element. Once at that element, you can manipulate it to create your desired effect. The DOM adds interactivity to every element on the page.

Ajax

Ajax, like DHTML, is an amalgamation of technologies. Created in 2005, *Asynchronous JavaScript and XML—Ajax* for short—combines CSS, HTML, DOM, XML, and XMLHttpRequest to allow a web page to perform functions in the background. Let's break this apart. Imagine MyAppoly has a form for individuals to complete that requires the user to enter their state. Most modern websites will suggest a few options for you in a dynamic drop-down menu, so when you type in "M," a list of states beginning with M automatically appears beneath the search box. When you type in an A, Massachusetts appears for you to select. Without Ajax, the alternative would be a page that refreshed entirely every time you typed a new letter so that it could get a new list of potential states that might be valid. You would type a letter, which the page would take and use to search through its database. After it found all states satisfying that letter, the page would refresh to display the updated information. Assuming it takes 0.5 seconds for a page to refresh, people would become impatient and annoyed with your website—a bad user experience.

So what does Ajax do? Instead of refreshing the page with the new content, Ajax instead transfers data the user requests from the server to the client. Ajax uses XML to format this data, and XHTML, CSS, DOM, and JavaScript are then used to take the data and update only the portion of the web page that is affected. The major point is that this is all done asynchronously, meaning the user does not know it is happening in the background. In fact, the user can continue to interact with the web page while Ajax is working to get more up-to-date information from the server. Ajax is sophisticated and often difficult to use, but many toolkits make Ajax much simpler.

Ajax only works in modern browsers, and so this technology too lacks backward compatibility.

Portability and Accessibility

There is no enforced standard for browser rendering, so all browsers render HTML and CSS slightly differently. What appears fine on Safari may look odd on Internet Explorer. What looks okay on the PC version of a browser might be unfit on a Mac. Due to changing technologies, what works on a version of Chrome today might not work on an outdated version. All of these observations mean that, as the developer, you are not reaching your entire audience in the same way. It is in your interest to either make sure your website looks beautiful on every display of every potential device—which is not realistic—or hope that web standards or some other solution comes along.

Web Standards

Imagine if every car company required a unique type of gas. It would be nearly impossible for gas stations to carry every type of gas a customer might need. Auto manufacturers realized from the outset that designing their vehicles to run on only a few industry-standard grades of gasoline was the optimal solution.

Currently, the Internet is like a world without gasoline standards. A much better solution to the issues of accessibility and portability is through web standards for both coding and rendering. The W3C is "an international community where member organizations, a full-time staff, and the public work together to develop Web standards." It produces and promotes not strict rules but rather suggested guidelines. Many note a convergence toward these recommendations recently, something that would be extremely beneficial for developers and users alike.

Several validators exist to ensure that your web pages satisfy the guidelines of the W3C. Validation is beneficial because it helps to make sure your web page is most widely accessible, often improves display time, and might help with search optimization (discussed in Chapter 10).

Responsive Design

What if you could spot which device people were using when they came to your site? If that were possible, you could then show them a special version of the site adjusted for their device. This technique is known as *responsive design*. Because it is possible to identify where a user is coming from and what device he or she is on, you can choose to display a certain CSS sheet or use different versions of the code that make the experience more optimal for the user. However, as you can imagine, covering all browser and device combinations is difficult.

Conclusion

You now understand what is required conceptually to translate the paper version of MyAppoly into a working front end. At this point, you tell your CTO that you understand how users access a website, interact with your functionality, and which programming language processes users' actions. But how does your back end remember and organize information, such as a user's name or their list of friends? To answer that question, you need to understand databases, the subject of the next chapter.

Databases: The Model

The information that some companies collect is an asset in and of itself due to the value of data. Working for a startup that specialized in providing data to local businesses, I focused on collecting menus from local businesses, primarily restaurants. With a database of all food items and their prices sold at restaurants, I was able to learn some interesting facts! What is the average price of a pizza in San Francisco? Which restaurants within a five-mile radius of my hotel have kimchi on the menu? No problem. Collecting unique data might allow you to answer previously unanswerable questions.

The last component of a "typical" application that you need to familiarize yourself with is the *database*, often called the *model*. Users of MyAppoly will need to enter their information in order to sign up for your service. You have to store their data somewhere. From usernames and passwords to the number of user logins, the amount of information that you will want to save will grow exponentially as your site becomes more and more successful. This chapter will prepare you to join your team's discussion of what type of database to use and how to organize your data. While this chapter does not discuss how to add and update data in a database programmatically, it reviews how databases are typically structured and how they differ.

© Vinay Trivedi 2019
V. Trivedi, *How to Speak Tech*, https://doi.org/10.1007/978-1-4842-4324-4_5

Database Systems

Most of us are familiar with some sort of record-keeping system. Some people store their receipts in preparation for paying taxes. Business owners might keep track of their sales and expenses to make sure they are hitting their targets. Physicians keep files on all of their patients, typically organized by last name. A database system is an electronic version of a filing cabinet, and "requests" are interactions with the cabinet, including adding, removing, updating, or retrieving a file, or *record*.

Which type of data might you collect? Typically, you want to save data that will be of some value to you in the future. This is called *persistent data* because it is saved beyond a day or after a user logs off. Because the power of data and analysis unlocks efficiency, insights, and opportunities, databases are critical to all businesses today.

The Four Components of Databases

Databases consist of four components: data, hardware, software, and users.

Data

Data is the sine qua non of a database—the actual information saved in the database. In addition to the data, databases also store meta-information about the data and the database. For example, records are comprised of multiple pieces of information, so it is useful to know which part of the record corresponds to the username, as an example. This *metadata* is saved in the *catalog*—just in case you need to refresh your memory on what you are collecting and how it's being stored.

Hardware

Data is physically stored on some component of hardware—typically in secondary storage, such as magnetic disks. Storage hardware can be categorized in several ways: primary, secondary, and tertiary.

Primary storage provides fast access to data because the computer's central processing unit (CPU) directly interacts with it. The CPU carries out the instructions of a computer program, so think of it as the brain of the process or the executor of the program. When a computer performs a task that you requested, it first accesses primary storage for the specific instructions of what to do before the CPU executes. Terms associated with primary storage include registers, cache memory (also referred to as static random-access memory or RAM), and dynamic RAM (DRAM) or main memory. These various parts are associated with different tasks, but they all serve to keep data and programs that the CPU is actively using ready for execution. We say that primary storage is volatile because after the machine is turned off or is without power, the contents are lost. Primary storage is costly, however, and therefore tends to be limited in size.

Secondary storage, by contrast, is nonvolatile, cheaper, and not directly accessed by the CPU. Because the data persists even after the computer is turned off and because we typically would like the safety of our data to be independent of our computer's power, it may not be surprising to learn that secondary storage serves as our mass storage site. The CPU cannot directly access the contents on secondary storage; data must first be copied into primary storage before execution, a fact that highlights why accessing data in secondary storage tends to be slower. Hard disk drives, flash memory, and floppy disks are examples of secondary storage.

Tertiary storage is similar to secondary storage in that it is not accessed directly by the CPU. It is used for massive data storage and is much slower than secondary storage.

The key trade-off distinguishing these three classes of storage is access time vs. expense, a consideration touched on in the "Optimization" section at the end of this chapter.

Software

When users interact with MyAppoly, they will perform certain actions that you will want to save in your database. A *database management system* (DBMS) is responsible for defining the database and storing, manipulating, and sharing the data across users and applications.

Think of DBMS as the head of file management, responsible for managing the requests of all of MyAppoly's visitors. When a user clicks "accept friend" on Facebook, a request is created and directed at the database. The DBMS accepts and analyzes all requests before sending them along to the database, at which point they are *interpreted* (converted into a request the database can understand) and executed. The code that actually accesses the database on behalf of the DBMS is called a data sublanguage; one popular version is SQL.

The DBMS provides a way to control for *concurrency* (multiple requests to the database happening at the same time), *redundancy* (the same piece of information stored multiple times in the database), security, *recovery* (protection against some sort of database failure), and *optimization* (timely handling of requests)—aspects that greatly improve the value to the developer.

Users

Your end users are the users of MyAppoly who, by interacting with the application, generate requests to the database in the background. When a new user registers, a database request is issued asking the DBMS to add a new user to the database.

Three-Level Architecture

You will also hear database systems described as an *architecture* consisting of three parts: the physical, the logical or conceptual, and the external. The physical layer is where the data is stored, the logical consists of how the DBMS interfaces with the application, and the external layer includes user interactions with the database.

The three levels, three tiers, or three schemata of database systems are similar to our earlier database description, but it is helpful to understand jargon that might be thrown around in the context of database systems. It is also now easier to explain the database independence principle. The DBMS provides a form of abstraction so that users can interact with the database without needing to know the details of how the data is stored. As a result, any of the tiers can be changed in its implementation without affecting the entire system. Let's say your programmer codes the application in a way that displays all of the names of MyAppoly users in a list. You would like all the users, however, to be displayed on a graph. It would be a pain for the programmer to rewrite requests to the database to get the information.

Thanks to the data independence principle, your programmer can change the list to a graph, and it will not affect how the programmer interacts with the rest of the system. The change only affects the visualization layer and has nothing to do with how you get the data or how the data is stored. In summary, the system separates the management of data from the actual use of the data, which has several benefits. One could argue that the process of writing requests that get handled by a middleman (DBMS) before going to the database slows down the processing time, but the benefits outweigh the lag cost.

Classification

Criteria for classifying database management systems include the following, which are considered in turn in the succeeding sections:

- *Data model*: The data need to be stored in a way that accurately and efficiently describes the real world. Different types of models are discussed in the next section.

- *Number of sites:* *Centralized* and *distributed* are buzzwords you have probably encountered before. How they relate to databases is discussed in the section "Centralized vs. Distributed."

- *Cost:* Solutions can run the cost gamut. Free and open-source solutions, such as MySQL, are available. Commercial solutions are offered to larger enterprises that want more customizable systems.

Data Model

There are several ways to store data in a database that are efficient and organized. Four of the most common data models are discussed in the following sections: the relational model, the non-relational model, the object-oriented model, and the object-relational model.

Relational Model

A model proposed in 1969 by Edgar F. Codd called the *relational model* became and continues to be the de facto standard. Relation is nothing more than a mathematical term for a table, so from that fact alone, we can ascertain that the relational model uses tables to represent data.

In the relational model, which is also known as the entity-relationship model, everything that exists, such as a car, a class, or a store, is an entity. Each entity has attributes, which are the columns of the table: a car has a color, a class has a name, and a store has a location. Each row of the table represents a specific entity (also can be called an object or record). In our users table, each row will describe a user of MyAppoly. The columns can be items such as first name, last name, password, address, and email address. It is common for tables to have a primary key, or a column for which the value is unique for every row, usually a unique ID for that entity. A unique user ID in the MyAppoly table would allow us to refer to individual customers with no chance of overlap. Most websites require usernames to be unique, meaning usernames could be used as the primary key instead of a user ID.

Now that the database models our entities and their relationships well, how do we interact with the database? Typically the database and its specific tables are modified by using particular operators that the DBMS and its data sublanguage, the language used to communicate with the DBMS, provide. As with all languages, the sublanguage is picky and requires requests to be formatted in a specific way, so you have to spell out very clearly what you want. For example, *Structured Query Language* (SQL) is a proprietary query language (data sublanguage) developed by IBM that is now the de facto international standard and supported by most relational systems. In SQL, these operators would include INSERT (add entry to table), DELETE (remove row from table), SELECT (retrieve information already stored in database), and UPDATE (modify existing rows).

There are several commands worth knowing that allow for more complicated operations. One example is JOIN, which is used in conjunction with SELECT. Let's say you have the user table as described above and a separate table called pets that stores data on all of the users' pets. Every row of the pets table represents a pet, and the columns include name, owner, color, and type. In this example, the owner column of the pets table should correspond to a user in the user table. What if I wanted to answer the following question: "What are the addresses of all users who own pythons as pets?" This request involves accessing the pets table to retrieve all owners who have pythons and then giving each owner in your list of owners to the users table to obtain their addresses. This two-step process can be simplified by using the JOIN operation, which essentially combines two tables based on a shared column. (In this example, the shared columns would be user and owner.) This example illustrates that requests can get complicated given the structure of relational systems, but operations exist to let developers obtain answers to their questions.

The query language and operators just described are nonprocedural, meaning that users can specify what they want but not necessarily how the DBMS should go about executing the request. I'm not sure if you would have a strong opinion on how specifically your data is retrieved, but it could matter from a speed perspective.

Let's review a few other concepts and terms for thoroughness. A series of requests that make up a particular task is called a *transaction*—all of the requests required to register a user and his or her pet, for example, could define a transaction. A few rules exist for transactions that are not exclusive to relational systems but will be discussed here briefly. First, transactions are *atomic*, meaning they are all or nothing—either the user is entirely registered or the database remains unchanged. As a result, even if the process were to fail halfway through for whatever reason, the table is still consistent, meaning there are no incomplete rows or partially updated records. Second, transactions are *durable*, meaning that they immediately change the database and the changes are available right after the transaction execution completes. Any lag in transactions being reflected in the database would negatively affect the entire experience of the site; imagine if it took a week for your Facebook friend list to update every time you accepted a new friend. Third, transactions are *isolated*, meaning two transactions are considered independently by the DBMS. Designing a database that adheres to all these constraints is difficult, but these descriptions should help you understand how a DBMS thinks about requests.

One last term to mention is *normalization*, which refers to the process of eliminating redundancy in databases. Let us go back to our example of users and pets tables. What if a developer created a column in our pets' table called owner's address? Isn't that information already saved in the users table? In this case, the address columns would be redundant and waste space. While normalization can help database users conserve memory, it can also take longer to get the results for queries. If I frequently ask the question I posed earlier about the address of all people who own pythons (the one for which I had to use the JOIN), my stance might be different. JOINs can be expensive because the DBMS has to combine two tables and then find the information, a potentially costly process depending on the size of the tables. It might just be easier and more economical to create the redundant column. These types of decisions can be made by *database administrators* (DBAs), who understand precisely how requests are executed.

Non-relational Model

Non-relational models, as you may have guessed, do not use tables as the primary data structure. Therefore, this category encompasses a lot, including networks, hierarchic models (think family trees), and lists.

One term worth a special mention is NoSQL. Relational systems are suited for more complicated data where queries will be requested that concern the relationships between entities (e.g., between users and their pets). In contrast, NoSQL systems tend to be great for storing large amounts of simple data such as key-value pairs (firstName:Vinay). For these simple use cases, NoSQL offers great performance and scalability. If your team feels that such a strict relational structure is not necessary and that you are dealing with storing and retrieving large amounts of data where the relationship between stored items is not as important, NoSQL database management systems are very useful and arguably more effective. Open-source NoSQL databases, such as MongoDB and Cassandra, are worth exploring.

Object-Oriented Model

Object-oriented databases center around the concept of the *object* (Chapter 3) in the context of object-oriented programming languages. These objects describe entities, and each object consists of a series of properties or attributes describing it, as well as a series of actions that can be performed on each object (e.g., objects describing pet owners may be able to purchase a pet). As before, there has to be something unique about each object, so an Object ID is often used. You can think of encoding relationships between objects in the following way: the object describing a pet owner might store the Object ID of the pet object that describes the details of the pet. So if you want the details of user X's pet, you could access user X's object, get the user's corresponding pet ID, and then access the pet object. These objects can sit in a hierarchy to give more structure to the system and take advantage of inheritance and other qualities associated with object-oriented models.

Object-oriented models are powerful because they allow you to preserve the complicated nature of certain objects. In addition, they help separate the structure of objects from the operations you can apply to the objects. The details of how these models work are left for more technical works, but many of the terms discussed under "Relational Model" carry over to this one. For example, object-oriented models also have a query language: *Object Query Language* (OQL).

Object-Relational Model

Object-oriented aspects have found their way into relational systems to merge the best part of both worlds in *object-relational models*. In an object-relational system, you can define your own type, for example.

XML

As mentioned in Chapters 3 and 4, XML is a markup language that can be used to give structure to documents. Therefore, it can be used to define the structure of data and is a primary way data are communicated over the Web.

Centralized vs. Distributed

The client-server architecture describes the relationship between the end user and the DBMS. Clients consist of all the applications that interact with the DBMS, and the DBMS is the server that processes and executes the requests.

The actual computer that stores everything can either be *centralized* or *distributed* in nature. When centralized, all requests come to a single server that houses all of the data. If this server were to go down for whatever reason, the application relying on it would not work, making the server a single point of failure. Therefore, as communication and database technologies improved, organizations realized there could be many benefits to a distributed system. Rather than having one central database server, what if we had multiple that could communicate with each other? In a distributed system, different parts of a single request could be executed on different machines within the distributed network of database servers. Though the data may rest in several sites all managed across a network by software, to the users of the application, the database would function as if it were centralized. To revisit our example, the users table and the pets table could theoretically be stored on different servers.

A typical distributed system has several qualities, but some of the most important ones include local autonomy (every site should be independent of one another), no central site, local independence (users should not need to know where the data is stored), and pretty much any other property that allows it to function exactly like a centralized system.

As is evident from the popularity of the word *distributed* in the technical world, distributed processing has become popular, and for good reasons. Benefits include the following:

- *Superior rationality:* If you think about the typical organization, data already tends to be distributed. Whether it is spread across geographic offices or internally by division, databases have logical divisions. Therefore, if the other benefits are compelling, why not retain this distributed quality? It could allow each group to manage their data and their data alone without having to come up with a way to combine it all into a centralized server. It might make sense for Pepsi Asia, for example, to store some of its data in Asia.

- *Improved efficiency and performance:* Intuitively, the processing of data is most efficient when done closest to the site where the request was given. Therefore, data should be stored near the people accessing it the most often. A distributed system can help with the most efficient allocation of an entire organization's data. In addition, performance can be improved when queries are spread over multiple machines.

- *Increased reliability:* Because information is available on various computers (several copies can be created as well), the probability that the entire system is down at any given point is much lower than in centralized systems.

- *Adjustable expansion:* Distributed systems are horizontally scalable, which means it is easy to add more computers to the system. Say for MyAppoly you are currently using five computers around the United States to store information. It is much easier to add a sixth computer than it is to vertically scale, i.e., adding more memory to a particular computer. Services such as Amazon Web Services (Chapter 2) are so popular because they allow organizations to add and take away servers depending on demand. This seamless way to scale and shrink allows organizations to spend their resources more effectively.

No sense in buying a lot of capacity when you only ever use that much capacity once in a while (e.g., when you have a promotion on your website).

The drawbacks to such a system are fewer, but they exist. First, because data can reside in several sites, query processing that requires bringing together data across servers takes more time. Second, the various functionalities that the DBMS provides (security, concurrency management, and so on) tend to be more technically complicated in a distributed system compared to a centralized system.

Other Topics

Other database management requirements concern concurrency, security, and optimization, treated in turn in the next sections.

Concurrency

You do not want multiple database users to have conflicting requests. If a Facebook friend blocks you at the same time you send them a message, how will Facebook handle this? This question relates to *concurrency*.

In the context of databases, *concurrent access* means that multiple users can access the database or multiple transactions can be processed at the same time. Because DBMSs support concurrency, they also need to be able to control the potential dilemmas that could arise when multiple transactions are trying to alter the database simultaneously.

For example, how does Yelp handle two users adding a recommendation to a restaurant on Yelp at the same time? Knowing that DBMSs manage this is enough, but for more specifics, a few concurrency problems follow:

- *Lost update:* When two transactions attempt to update the same value, one might be lost.

- *Uncommitted dependency:* One transaction can access a value that was updated but not yet finalized by another transaction.

- *Inconsistent analysis:* When two transactions have competing objectives, the steps might not be consistent with each other.

These problems are typically solved with a concept known as *locking,* whereby a particular object or value will be locked and accessible only by one process. Locking adds wait time between each process to ensure no two transactions collide or interfere with each other.

Security

Database security breaches are happening more regularly. Customers of any site demand a level of protection, and companies must ensure that their data is kept secure.

DBMSs protect data against unauthorized access and improper modification of data (preserving data *integrity*) in the following two ways:

- *Discretionary control*: In this model, every user is assigned appropriate access rights to different objects in the database. Discretionary control tends to be the more flexible of the two control types.

- *Mandatory control*: With mandatory control, every object in the database is assigned a classification level and every user is given a clearance level. Users can only access the set of data objects in their classification level.

Essentially, the DBMS can label users and data accordingly to make sure that no unauthorized person can access or manipulate private data. As added checks, the DBMS logs the transaction, time of transaction, and user who issued the request. These logs can be used to track suspicious activity (see Chapter 12).

Several complications can arise that do not immediately seem solvable by the control systems described here, although DBAs can use DBMSs and other tools to secure against them. One example pertains to the flow of information from the "haves" to the "have nots." A situation could arise in which a user with higher clearance, for example, alters the data in a way to give someone with lower clearance access to it. This flow must be controlled.

Another issue relates to *interference*, a problem for statistical databases. MyAppoly might use statistical databases that do not provide users with specific information about other users, but do offer aggregate statistics about a pool of users. Individual users reluctant to allow others to know particulars about their salary, age, or some other personal metric might be more willing to share information if it is only accessible when made anonymous in the form of a statistic. One problem is that particular filtering might allow someone to infer something about a particular user, something that should not be allowed. For example, if you request the average salary of all python owners in Cambridge, but there is only one such person that fits that description, the average salary that the request will return will be an individual's salary. DBMSs should protect against this.

Data encryption is a method used for protecting data stored or communicated across the Internet (Chapter 12). Important or sensitive information, such as Social Security numbers or credit card information, should always be protected by both restricting access and encrypting the information.

In summary, DBMSs and other tools exist to protect data. No solution is perfect, which is why security is at risk. The goal of a security system is to deter incursions by rendering the cost of breaking into the system greater than the potential payoff.

Optimization

There are numerous ways in which database systems can be optimized to use less space and to return or update information more quickly. One way is to design more efficient queries. For example, if you wanted to know how many individuals lived in California and owned a python, it's faster to search through a couple of thousand python owners for California residents than to look through millions of California residents for python owners.

Another area in which databases can be optimized, especially for space, is redundancy. Data is said to be integrated when redundancy among information is removed. It is wasteful, for example, to have a patient's name, insurance plan, and address written on every piece of paper within his or her file. Integrated data implies that such repetition is removed and kept in only one table in the case of relational databases.

Above most other considerations, database design must be intelligent. It should reflect the things you need the fastest or update the most. Otherwise, regardless of how well queries are written or what tools are used, a flawed database design will be suboptimal and result in wasted time, money, and space. You should only store the information you need, and you should organize it in a logical way that keeps related information together. The way in which you store your information should be determined by how you expect to use it later.

Big Data

How is *big data* different from normal data? Organizations have of course been amassing information even before there were computers, but the amount of data that can be collected and stored has dramatically increased. The size of data collected today is unable to be processed by our traditional methods. Relational databases are being pushed to the wayside because they are simply too costly and restrictive; the more scalable and flexible non-relational models are coming to the fore. To give you an idea of scale and growth, it is estimated that 1.2 zettabytes (1.2 × 10^{21} bytes) of data were created in 2010.[1] As a comparison, it is estimated that there are between 10^{20}

---

[1] Jim Kaskade, "Making Sense of Big Data (webinar video)," 2010, http://www.infochimps.com/video/making-sense-of-big-data/.

and 10^{24} grains of sand on our planet. Imagine sifting through all of that! If you assume the amount of data we have doubles every couple of years, we are in store for a lot of data centers.

Before commenting on how one should even begin to study so much data, there are a few terms used to describe data. The first is *volume*, one that is most obvious from big data's name. The amount of information is the largest problem and has prompted engineers to think about how distributed systems can serve as the solution. Next is *velocity*, which refers to the rate at which new data is created. The velocity of Facebook data is probably much higher than that of MyAppoly, for example. With so many interactions on Facebook every day, Facebook needs to make an important decision: what do they keep and what do they throw away? Online shopping websites face a similar dilemma. Every time a user clicks an item or writes a search term, the company must decide if they want to save that information. Saving more information, however, may compromise the speed with which they are able to convert data into insights. If retailers are slow to customize the web experience for their customers due to a delay in data processing, faster competitors are likely to win. Last, there is *variety*. If the data comes from several sources and in several forms, it presents a much more complicated set of issues to tackle.

You can now go to your data team and tell them that the variety of your data complicates your ability to store it and that the current velocity is overwhelming your limited volume server capacity. More colloquially, you might just say "We need more money!"

Let us assume you have a large volume of data of significant variety. The chances that you can make any sense of the information by looking at it are little to none, especially if there is no obvious pattern. One word you might come across in big data is *Hadoop*, a software framework that allows organizations to run tests on their data with the hope of answering questions. As covered in earlier chapters, distributed architecture is the key to our ability to store and process this quantity of information. Hadoop allows organizations to make sense of their data by spreading it out over a bunch of powerful processors that work in parallel to study the information. The general method Hadoop uses is called *MapReduce*, which consists of two parts: map and reduce. Let us say you have some question you want to determine from your dataset and your engineers write a function that theoretically will yield the right answer. *Map* will separate the task into several parts and distribute it across several machines. When this phase is complete, *reduce* will combine the results from each machine into a single task to complete the analysis. No single machine would be able to do this much work in a timely manner, so Hadoop's divide-and-conquer strategy is a very powerful tool. Other tools, such as *Spark*, are also gaining popularity.

Conclusion

At this point, you have learned some of the basics of how MyAppoly runs on the back end, what users will interact with at the front end, and the way data factors into all of it.

Leveraging Existing Code: APIs, Libraries, and Open-Source Projects

Imagine how much more slowly society would have developed if no one could leverage the inventions of others to design new products. Using others' innovations for new discoveries is crucial. Maybe a drug developed for one disease can be adapted to cure another. Perhaps flexible glass created for more durable dishware can be used to make unbreakable cell phone screens. We continuously search for the next best thing, and we cannot expect to find it if we always begin from scratch.

© Vinay Trivedi 2019

V. Trivedi, *How to Speak Tech*, https://doi.org/10.1007/978-1-4842-4324-4_6

Luckily, MyAppoly doesn't need to start from scratch either. There are a myriad of ways MyAppoly can leverage existing code. Say you want to include a map in MyAppoly. Building the map from scratch is hardly worthwhile for your team, given that other companies have already dedicated the time and money to do just that. Google's willingness to allow developers to use Google Maps points to a vital driver of the rapid proliferation of websites and applications in the Web 2.0 era. The popular adage "Don't reinvent the wheel" defines the tech world, which has developed a culture of sharing information. Internet companies carefully protect their information, but in general, they are more open to sharing their discoveries than organizations of the past. This collaborative culture partly explains the extraordinary growth, fertility, and wealth of Silicon Valley.

You want to focus on what makes MyAppoly different, not on rebuilding the code that other companies already offer to you. But what are these ready-made tools? This chapter describes the essentials of *application programming interfaces* (APIs).[1] It also addresses the open-source movement and its importance.

Application Programming Interfaces

Say you want to show your users the locations of the nearest ATMs on a map and accept credit card payments on MyAppoly. Either your team can build these features from scratch, which will increase development time considerably, or you can utilize the work of others, such as Google Maps and Stripe. Some APIs are free, while others require payment. Google Maps allows you to use their maps within MyAppoly for free, whereas Stripe collects a minor fee for each credit card payment they process for you. The ability to easily integrate technology by plugging one company's product into another's has allowed startups to grow quickly and become sophisticated in a relatively short time. MyAppoly can benefit from not only the APIs of others but also by releasing one of its own. If other companies become dependent on the data and functionality made available through MyAppoly's API, your business will become even more attractive.

---

[1] This chapter draws adaptively on schemata in Paul Reinheimer, *Professional Web APIs with PHP*. Wrox, 2006.

Using Others' APIs

Notable advantages and disadvantages of using APIs include the following:

Advantages:

- *Comparative advantage*: Using others' APIs allows you to focus on your core competency and value proposition. In the same way a traditional business might contract out the nonessential work, such as document editing or reimbursements, you can contract out such work as payment processing and map features.

- *Save development time*: Contracting out your noncore work to other companies by using their APIs saves your development team's time. You will save time both in building the feature and in maintaining it.

- *Leverage the expertise of others*: Companies typically release APIs in the area of their core competency. Whether their API provides information or functionality, chances are the company invests a lot of time into developing it. Even if your team could do a great job, why not give the responsibility to a company that is purely focusing on solving that problem?

- *Access to information*: Many APIs provide access to data. Often these data are available on the company's website but in a format that is not conducive to collecting at scale. An API can give you the same data in a format that allows you to manipulate it more easily. An example could be a weather API. The website might show you the daily weather in a particular city over time, but imagine that you were conducting a global warming study and had to collect the daily data from ten cities over the past 50 years. It would be tedious to copy each number from the website manually. Their API might allow you to access and download all of the data you want at once.

Disadvantages:

- *Dependence*: If the company behind an API decides to change its policies or shut it down, your web application will no longer function properly. This risk is a part of any business that depends on third parties. Do they have a lot of big clients? Are they performing well? Has their API been available for a while? Answers to these questions help predict reliability.

- *Not custom-made*: It is also possible that your exact needs are not satisfied by any API on the market. APIs provide general solutions, but they may not work perfectly for your site. Therefore, there may be a tension between conforming to what an API offers in the name of the advantages listed above and building the entire feature from scratch to get the full range of functionality.

Where you come out after weighing this list of pros and cons will depend on the specific API and its functionalities. Just remember that leveraging existing code is a trend across the industry, and excessive conservatism may drop you behind MyAppoly's competitors.

Making an API Available

Imagine that MyAppoly allowed its users to match random handwriting samples to specific individuals. You might be building this tool for forensic investigation, electronic payment verification, or historians. One way you could enable all of these scenarios would be to release a MyAppoly API. The advantages of releasing a MyAppoly API include the following:

- *Scale client integration*: If you want to give different potential partners access to your data or functionality, building a custom solution for each one would require a lot of time and resources. Why not standardize the interface in the form of an API and let your partners access what they want within the parameters you provide?

- *Others' dependence on you*: If your API becomes a critical part of the operation of several companies, you have generated an additional source of value for your company. Others rely on you. Think about all of the third-party applications built on top of Facebook. If Facebook disappeared, imagine how many other companies would collapse. Other companies' reliance on Facebook undoubtedly supports Facebook's valuation and success.

- *Variety of distribution methods*: An API can make the same information that you already present through your application available to people who want to consume it differently. By providing this alternate access method, you are increasing your market.

- *Revenue source*: Charging others for using your API could represent a substantial source of revenue.

- *More data*: Every time someone makes a call to your API, you can record that information. This data might help you identify trends in the industry or opportunities for new features.

- *Brand value*: Even if you choose to make your API available for free, the value you generate from widespread use will return to you in the form of brand and reputation. It can help establish you as a leader in your field and a proponent of knowledge sharing.

You must also consider the disadvantages:

- *Resources*: APIs bring additional engineering and maintenance challenges. Every time your data source changes, you must make sure the API works properly. For example, if you expand your data set to include audio files, you have to update your API correspondingly. Lastly, developers trying to integrate your API into their applications might require customer support and documentation explaining the available functionality. An API without support will lead to frustration and eventual abandonment. If you want to make an API available, be forewarned that it can be a resource-intensive process.

- *Security*: An API allows users to gain access to your data. This portal might expose vulnerabilities whereby other information you do not want to share can be improperly accessed. The proliferation of APIs suggests, however, that these risks are avoidable or manageable.

- *Cost*: Every time that a developer uses your API (by making an API "call" or "request"), MyAppoly must take action. Who pays to process these requests? You do! Before you release an API, you need a plan to pay for handling the requests.

The success of your API will depend on how easy it is for developers to learn, use, and integrate.

How Do APIs Work?

The technical details underlying APIs can get complicated, but the basics are simple. APIs are software-to-software interfaces. Applications talk to each other without your involvement. If MyAppoly needs to show the weather, for example, the software can communicate ("make calls") with a weather API to obtain the most updated weather report before showing it to the user.

One way you could retrieve the weather would be through the weather website (direct access). Alternatively, you could go through an API. Instead of accessing the weather directly through a browser, you write code that uses the API to get the information you want. But how do you know what to write? The API provides *documentation* that tells you what the valid requests are and how exactly to write them, similar to a programming language. So, after reading the API's instructions and rules, you specify that you want the weather by using its *getWeather* function. When you give the function to the API, it interacts with the server to retrieve the information you requested. The API then returns the results to you in a format that is easily digestible (typically XML or JSON, discussed in the "JavaScript Object Notation" section of this chapter). Depending on what the API allows you to do, you could theoretically get all the information you want through the API and never have to visit the website again through a browser.

From this example, you can see that APIs are designed for software developers and can be defined as "a set of routines (usually functions) and accompanying protocols for using these functions that provide building blocks for software development." They allow users to connect to the application via a secure channel and then run functions using code to get information or to borrow functionality.

Two of the more popular types of APIs—*Representational State Transfer* (REST) and *Simple Object Access Protocol* (SOAP)—are described and compared in the next sections.

REST

REST is an architectural style developed by Roy Thomas Fielding to allow the Web to be more scalable. Every request has two parts: the endpoint, or URL, and the message that contains the request. This message consists of a few parts, including a developer ID or "key" if required (this helps the API keep track of who is making requests and how often), the desired action (e.g., *getWeather*), and parameters that give more information on the request (e.g., *getWeather* for today and yesterday only). All of this information is encoded in the URL and sent to the API via the HTTP GET method. The API then decodes the request, interacts with the server to complete the desired action, and then returns a response to the user.

SOAP

Started as a Microsoft initiative in 1997, SOAP is another way to request data from remote servers using HTTP. While the REST API encodes the specifics of the request in the URL, SOAP sends the details of the request in an XML document. The XML document that makes up the message has to satisfy the specifications outlined in a *Web Services Description Language* (WSDL) file. In a way, SOAP is like sending a request in an envelope (hiding the request's details), whereas REST is like sending a request on a postcard (viewable by all). Both travel via mail, or HTTP, though.

Comparing REST and SOAP

REST and SOAP present the following points of comparison:

- *Overhead*: Because REST requests are confined to a URL whereas a whole document describes SOAP requests, the latter tend to be larger, require more overhead, and contain more information in different parts of the message that further define the request.

- *Transparency*: To continue the postcard and envelope analogy, everything is out in the open with REST and hence monitored more easily, whereas only the address (endpoint URL) is viewable with SOAP.

- *Development time*: REST APIs are typically quicker to develop than SOAP APIs, contingent on your individual programmer's expertise.

- *Flexibility*: Because SOAP uses XML documents as messages, more information can be included in SOAP requests than in REST requests, giving SOAP APIs greater flexibility.

Authentication

As discussed in the opening sections of this chapter, you might want to keep track of who is using your API and what they are requesting. You also might want to communicate with your API users to confirm the accuracy of the data and security of the connection. Approaches to authentication include the following:

- *Open API (no authentication)*: There are no barriers to use the API, the code can be distributed more freely, and you do not need to manage requests, assuming you are okay not tracking individual users. Open API is an attractive option if you are interested only in anonymous, aggregated data, not in specific individuals' actions or in controlling access to the API.

- *Message-based authentication:* Credentials such as a unique developer's key can be passed in the API request as part of the message. Alternatively, information can be passed in the HTTP headers of incoming requests.

- *SSL endpoint:* This method helps the client keep track of the server. After receiving a server certificate, the client can check to see if it changes. If it does, the information is likely coming from a source other than the server and therefore cannot be trusted. This authentication method prevents an attacker from impersonating the person you are expecting. Because the SSL endpoint method does not help identify the client, it is used in conjunction with one of the other methods listed here.

- *Client-side certificates:* You can configure your API to create certificates to give to a client (via a secure channel) to present for authentication every time the client makes a request. Although client-side certificates are considered a robust way of dealing with authentication, this form of authentication can be considerably slower.

JavaScript Object Notation

Another formatting language besides XML (Chapters 4 and 5) that you are likely to come across is *JavaScript Object Notation* (JSON), which uses a collection of name/value pairs that can be nested. For example, a book (object) can contain many name/value pairs that describe its author, title, and year published. Many APIs "return" (deliver the response in) JSON rather than XML. These are different ways of structuring the same information.

Libraries

A *library* is a collection of code that defines commonly used functionality (such as drop-down menus or animated pop-up windows). For example, you might create a library of math variables and functions. You can share your library so that anyone else who wants to use math in their programs can import your shared library and save a lot of time. Shared code and libraries reduce development time and the number of developers needed.

On the other hand, how much do you trust the reliability of your system on someone else's contributed software? There are no right or wrong answers, but your ultimate decision requires good judgment on behalf of your team of programmers.

Open Source

Related to the spread of information through APIs and libraries is the growth of *open-source* projects. *Open source* is best explained by examining the history of software development.

Though hard to imagine, there was a time when all software development resided with researchers and was developed at universities; the discoveries could be shared without any limitations. All research was for the public good, after all. During the 1960s, corporations such as IBM and others borrowed this custom; when they released their first large-scale commercial computers, they gave away their software for free. With developments in technology, the arrival of competition, and falling prices of hardware, companies saw software as an alternate revenue source that could support their business. Software became commercial, proprietary property, and any reproduction without the company's knowledge was illegal.

Richard Stallman at MIT was upset with this change in culture. Concerned that MIT would control his new project and its distribution, Stallman left the university to continue working on his GNU software, a computer operating system made of entirely free software (such as a free version of Windows or Mac OS). As Stallman developed, his opposition to commercialization and his advocacy for equal access inspired others. In 1985, he founded the Free Software Foundation (FSF) to support the GNU project. The FSF promulgated the General Public License (1988) to protect free use and free access under the doctrine of *copyleft*. Rather than protecting a property's rights (copyright), "copyleft protected the freedom to copy, distribute, and change a product."

The declaration of the GPL challenged the commercial view of proprietary software, and in the 1990s, a misunderstanding over the word *free* emerged. The "free" in FSF pertains to freedom, not to its price.

According to the FSF definition, "a program is free software if the program's users have the four essential freedoms:"[2]

- The freedom to run the program, for any purpose (freedom 0).

- The freedom to study how the program works and adapt it to your needs (freedom 1). As a result, users must have access to the source code.

- The freedom to redistribute copies so you can help your neighbor (freedom 2).

---

[2]"The Free Software Definition," www.gnu.org/philosophy/free-sw.html, 2013.

- The freedom to improve the program and release your improvements to the public, so that the whole community benefits (freedom 3). Again, ensuring this freedom requires access to the source code.

Through its public education efforts—FSF used such slogans as "free speech, not free beer"—FSF tried to correct the misperception that it was against commercial software.

Recall from Chapter 3 that the code that makes up a program is called the *source code*. To be read, understood, and altered, the source code must be accessible to the public—hence the term, *open source*. Access to source code, however, is a necessary but not a sufficient condition for a program to be considered *open source*. Anyone must be able to use the code in their own creations without restriction. Founded in 1988, the Open Source Initiative (OSI) promulgates "The Open Source Definition" as follows:[3]

Open source doesn't just mean access to the source code. The distribution terms of open-source software must comply with the following criteria:

1. *Free redistribution:* The license shall not restrict any party from selling or giving away the software as a component of an aggregate software distribution containing programs from several different sources. The license shall not require a royalty or other fee for such sale.

2. *Source code:* The program must include source code and must allow distribution in source code as well as compiled form. Where some form of a product is not distributed with source code, there must be a well-publicized means of obtaining the source code for no more than a reasonable reproduction cost, preferably downloading via the Internet without charge. The source code must be the preferred form in which a programmer would modify the program. Deliberate obfuscation of source code is not allowed.

3. *Derived works:* The license must allow modifications and derived works and must allow distribution under the same terms as the license of the original software.

---

[3]"The Open Source Definition," http://opensource.org/osd, n.d.

4. *The integrity of the author's source code*: The license may restrict the source code from being distributed in modified form only if the license allows the distribution of "patch files" with the source code that allow modification of the program as it's built and compiled. The license must explicitly permit distribution of software built from modified source code. The license may require derived works to carry a different name or version number from the original software.

5. *No discrimination against persons or groups*: The license must not discriminate against any person or group of persons.

6. *No discrimination against fields of endeavor*: The license must not restrict anyone from making use of the program in a specific field of endeavor. For example, it may not restrict the program from being used in a business or from being used for genetic research.

7. *Distribution of the license*: The rights attached to the program must apply to all to whom the program is redistributed without the need for execution of an additional license by those parties.

8. *The license must not be specific to a product*: The rights attached to the program must not depend on the program's being part of a particular software distribution. If the program is extracted from that distribution and used or distributed within the terms of the program's license, all parties to whom the program is redistributed should have the same rights as those that are granted in conjunction with the original software distribution.

9. *The license must not restrict other software*: The license must not place restrictions on other software that is distributed along with the licensed software. For example, the license must not insist that all other programs distributed on the same medium must be open-source software.

10. *The license must be technology-neutral*: No provision of the license may be predicated on any individual technology or style of interface.

Around this time, Red Hat was founded as an organization solely dedicated to improving and educating others on how to use Linux, an open-source computer operating system. It was the first company whose entire business model focused on copyleft products. Open-source systems began challenging the position of large software corporations, including Microsoft.[4]

Since then, many other open-source licenses have been created, such as the Mozilla Public License by Netscape. The growing culture of open source is represented today by about 50 specific, OSI-approved, open-source licenses divided broadly into the following two groups:

1. *Academic licenses*: Under these licenses, universities distribute their research to the public and allow their software and source code to be used, copied, modified, and distributed. The Berkeley Software Distribution (BSD) license is the archetype and represents the "no-strings-attached" approach. All code licensed under academic licenses gets added to a general pool of code that can be used for any purpose, including for the creation of commercial software. You can borrow code from this pool and never contribute.

2. *Reciprocal licenses*: Under these licenses, the code is entered into the pool of open-source software. Anybody can use the code from this pool for whatever purpose, but if you choose to spread your newly created code, it must be distributed under the same license. In other words, you may leverage this code to produce something new, but you must contribute your new creation to the pool of code for others' use. GPL is an example of a reciprocal license.

Many people have concerns about the quality of open-source code. One rubric for evaluation consists of reliability, performance, and total cost of ownership. Other metrics used to evaluate quality include portability, flexibility, and freedom. Whatever metrics you select, the key point is that the many advantages of leveraging existing code can be offset by the risk inherent in relying on others' code and systems. Nonetheless, several open-source projects have attained the same complexity levels and quality standards as commercial software.

---

[4]See, for example, Vinod Valloppillil, "Open Source Software," Microsoft Memorandum, August 11, 1998. Available at http://catb.org/~esr/halloween/halloween1.html#quote8.

SDKs

As a quick aside, you may hear the term SDK, or software development kit, thrown around the MyAppoly office, especially as the team talks about mobile app development (Chapter 13). To make it easy for developers to develop for a platform, like the iOS software powering Apple's iPhones, the platform provider might offer an SDK. This SDK can be downloaded and includes whatever a developer needs to start building apps for its platform. That might include documentation, an integrated development editor as discussed in Chapter 3, APIs and other helpful libraries, and related tools (e.g., for debugging—finding errors—or tracking performance). The goal is to make things easy, so the more building blocks that one can leverage reliably, the better.

Conclusion

With a more in-depth understanding of APIs and the other ways you can leverage existing code, your team sketches out a full development plan. The use of a few key APIs and shared libraries helps accelerate the schedule, but the team plans to slowly shift away from using one of them because it does not give them the full range of functionality they need for MyAppoly. They ruled out several other contributed libraries and tools because of their reported poor performance and unreliability. You must strike a balance between doing things in-house vs. outsourcing, and now that you have made the decision, your team can develop MyAppoly. But how will they work together to build it?

Software Development: Working in Teams

Too many cooks spoil the broth. This adage applies equally well to programming unless you organize your engineers in the right way.

You spent the last few weeks assembling a small team of ambitious engineers. They understand how the back end, front end, and database interact and are ready to start programming MyAppoly. Before they start, you want to come up with guidelines for collaborating. How will they work together? Can two of them work on the same file at the same time? How will one engineer understand what another engineer did?

Such questions are important for any technical endeavor. Various design principles and tools support team-based coding, making it faster and safer. To interact with your engineers, you need to understand the terminology of these principles and tools—particularly version control systems.

© Vinay Trivedi 2019

V. Trivedi, *How to Speak Tech*, https://doi.org/10.1007/978-1-4842-4324-4_7

Documentation

The simplest way for engineers to explain their code to their teammates is to comment their code. Every programming language allows programmers to write comments next to particular lines or blocks of code to explain how the code works or what it does. Example comments include "Changes the user's password" or "Adds user to the database." With these comments, new engineers can quickly learn the codebase, and programmers can effectively contribute to the code they are less familiar with. Comments also allow managers to review code more efficiently, as the English statements improve code readability. Commenting is one form of *documentation*, the set of descriptions and instructions for a process or feature. Documentation is usually in paragraph form to help engineers understand the design and implementation of a significant feature more quickly. If your head engineer for MyAppoly builds a new feature that allows the application to process credit card payments, they might want to write up a few paragraphs explaining how the feature is integrated with the rest of the program, how the code works, where the code is saved, how the logic works, what APIs they used, and so on.

In addition to these external notes, style of coding also matters greatly for teams. By *style,* I mean naming conventions (how you name variables in the program), use of indentation and whitespace, and structure of comments, among other things. Imagine code that asks a user for their name and then saves it in the database. One engineer might format the text and call the variable that stores the user's name "name," whereas another engineer might call it something opaque like "var293." One is obviously more easily understood than the other. These coding styles are important; as a result, detailing a standard style guide for all engineers to use is a common practice. Many programming languages also have their own style guides or conventions that can serve as a starting point. If MyAppoly engineers follow these conventions, all formatting and labeling will be consistent across features, and new engineers will be able to collaborate and review code more easily.

Program Architecture

There are several ways you can organize MyAppoly's code. In Chapters 3 through 5, you learned how code performs the program's logic, constructs what the user sees, and interacts with databases. If all this code were lumped together, the program would be slow and unmanageable: imagine if all of a program's code sat in one file. How could an engineer even determine where to edit in a file millions of lines long?

Thankfully, programming is a mature endeavor, and smart people have developed better ways to organize code. These design principles not only improve efficiency and code performance but also conduce to better collaboration and code maintainability.

One of the most common categories of organization is the multitier application architecture. The most widespread of these is the three-tiered architecture. The three tiers correspond more or less to the subjects of Chapters 3–5. One tier concerns all things connected with the database, often called the model (Chapter 5). The middle tier is the controller or application processing tier, which contains the logic of the program and relates to the back end of the program (Chapter 3). The last tier is all things that a user sees: the view or presentation (Chapter 4). There are several other architectures, but I'll focus on why these architectures exist in the first place rather than go into the details of how each one is different.

First, these architectures support parallel work (this design principle is often called *separation of concerns*). One engineer can work on the presentation of MyAppoly while another works on the model. The controller engineer can tell the model engineer, "I don't care how you do it, but I need the first name of the user." Thus, the view engineer and controller engineer can continue coding assuming they have some way to access the name of the user, and they don't have to worry about the code that physically goes into the database to get the name. The model engineer can provide the other two engineers with a simple description: "If you want the name, call the function *getUserName*. If you want the user's location, call *getUserLocation*." The model engineer can then code the database interaction in any way. From this example, you can see that the three parts of the application can be developed relatively independently because each engineer does not need to be concerned with the specifics of how the other parts are coded.

Second, this setup supports maintainability. Imagine that the controller engineer tells the model engineer they need a user's name, address, and ZIP code. The model engineer writes three functions to get these pieces of data from the database and gives them to the controller engineer, who can then manipulate them before passing them off to the view engineer to display to the user. Three months later, the engineers determine that the model code is too inefficient in getting information from the database. Luckily, the view and controller engineers do not have to change anything. The model engineer can update their functions to improve performance, and nothing else in the application has to change. If the team were not using a well-defined architecture, this database code could be in any file. It would be a lot harder to hunt down every database-related code and change it.

A third reason one might choose to use such an architecture is code reusability. Imagine if the view engineer wants to show the user's name on every page. The inefficient way of doing this would be to rewrite the database code that gets the user's name on every single web page of the whole site. Instead, the view and controller engineers can call the same function *getUserName* that goes into the model and executes the query to get the name. This abstraction enables engineers to be more efficient and economical with their code.

The sum of these three advantages is clear: shortened development time. Products can be built and maintained more quickly, and precious engineering time can be dedicated to more value-added activities.

Revision Control

The architecture of a program helps separate functionality so that multiple engineers can work on a given project at the same time. But just separating functionality is not a complete solution. How does a team of engineers effectively coordinate their programming efforts? How does one person notify the team that they have updated a particular file? Do they email the newly updated files? If you are an engineer about to update a file, how do you know that you are working on the most up-to-date version? Perhaps you introduced a bug or an error into the code, and you want to go back to an older version. How do all of these frustrations get resolved? *Revision control* (also known as *version control* or *source control*) refers to the management of changes to a document. Your team will likely use a revision control system, such as Git or Mercurial, to solve these problems. We'll discuss some of revision control systems' features and their advantages.

Conflict Resolution

The two dominant ways to help resolve conflicts when multiple engineers are collaborating on a given project are file locking and version merging.

File locking is a method of preventing concurrent access to a given file. Think of a standard document that contains an essay. If two people edit the essay simultaneously, how do we determine whose edits to include? Can they be merged? File locking gives "write" privileges (the ability to edit a file) to the first person who opens the file and locks the file for other users. If another user tries to open the locked file, it will only be available for viewing ("read" access).

The benefit of file locking is that it avoids complicated problems with merging two sets of edits to the same file. When Engineer 2 opens a file already being edited by Engineer 1, they will be notified. There are drawbacks, however, to this solution. Engineer 2 has to wait for Engineer 1 to finish editing, introducing a delay in the process and slowing down development. Imagine, in a worst-case scenario, that Engineer 1 called out sick for a few days after opening a file. Either Engineer 2 would be locked out of the file for several days or they would need to get a higher authority to force-unlock the file, a process called *file unlock*. Another drawback is that this system might give a false sense of security. Two files that depend on each other could be locked by different engineers and altered in such a way to compromise their compatibility.

The other dominant conflict resolution method, *version merging*, allows multiple users to edit a single file. After several engineers finish working on the same file, the system helps merge the changes. When the engineers all work on different parts of a particular file, merging is easy, just as one editor could change an introduction to an essay while another modifies the conclusion. In contrast, two engineers changing the same part of the file in different ways creates a *conflict* that has to be resolved manually. Though the term "conflict" sounds hectic, many feel this type of system works more smoothly because there is less wasted time. As long as engineers do not work on the same parts of files, few conflicts emerge, and the merging process moves seamlessly. The process works well for text-based documents, such as code or documentation, because text can be easily compared and merged. File locking is a better choice for music or art files that are not as easily compared.

Centralized vs. Distributed

Revision control systems are either centralized or distributed. With *centralized* systems, all of the code is stored on a central server, called the main *repository*. Whenever an engineer wants to open a file or save a file, they have to be connected to the network to access the central server. This model is referred to as a client-server model and is similar to the model for retrieving websites. Although a full copy of the code can exist on a client's, or engineer's, computer, the history of all the changes is kept on the main server. If for whatever reason the central server goes down, the code and its history are either lost or temporarily unavailable, a single source of failure.

In the distributed model, multiple copies of the entire codebase and history (repository) exist on the machines of the engineers who are working on the project. They can open and save files when not connected to a network and can perform functions more quickly. Each engineer has a local copy of the repository ("repo" for short) that is updated with changes, although a distributed system can also use a centralized server. Every local copy (or working copy) is in effect a backup. There has been a significant shift to more distributed models for these reasons.

Revision Control Advantages

Revision control allows engineers to submit code changes, and it keeps track of each change independently and sequentially. As a result, engineers can see a full timeline of the edits to a file and can quickly *revert* to a previous version of the file or even of the entire codebase. Moreover, each edit is accompanied by comments from its engineer explaining the change. Future engineers can easily see who was responsible for each change and ask for further clarity if needed.

Revision control systems are fast because they only track the changes, or *diffs*, to the files. For example, if you just added a sentence to a book, you could more quickly describe your edit by indicating where to insert the sentence and what to add instead of copying the entire new edition. Similarly, revision control systems track two things: the line numbers of changes and the actual changes (either added code or deleted code). Once saved, the diff is sent to the master version of the codebase, meaning other engineers will always have access to the most up-to-date version of the codebase.

Another significant advantage is that two engineers can use their own copies to work on separate features independently. If the features don't edit the same code, they are easily merged. Otherwise, the revision control system will have a conflict management system that makes it easy to combine the code for the two features or will use file locking to prevent conflicts in the first place.

Version control also allows engineers to *tag* the codebase in various states. Recall that revision control systems enable viewing of the codebase for any prior time. You may want to bookmark the codebase at certain times. For example, you may want to label the released codebase with a "version 1.0" tag. Similarly, you may want to tag well-tested but unreleased code as "beta" or the most recent changes as "latest" for easy access.

Conclusion

Programming in teams can present several obstacles, but with the proper style guide, program architecture, and version control system, teammates will be able to collaborate more effectively. Next time someone tells you too many cooks spoil the broth, tell them to use Git.

Software Development: The Process

In 2011, having spent the previous three years advising entrepreneurs on organizing resources and designing processes, Eric Ries laid out his methodology in the book *The Lean Startup*, the release of which sparked the Lean Startup movement.[1] Its philosophy, which articulates a continuous and feedback-driven approach to product development, informs this chapter.

Your CTO says that MyAppoly must have a well-defined process for developing the product. Should you aim to release a new feature once a month? Should someone build and test a new feature before showing it to the team? What should the flow of software development look like, and how can it be managed? This chapter does not directly relate to the technical aspects of a web application, but any process of developing software must account for the complexity and unpredictability of software production. A high-level understanding of how software development teams work will make you a better manager. Your team tells you about their design for the software development process (the way the product will be built). Without

---

[1]Eric Ries, *The Lean Startup: How Today's Entrepreneurs Use Continuous Innovation to Create Radically Successful Businesses*. Crown Business, 2011.

© Vinay Trivedi 2019
V. Trivedi, *How to Speak Tech*, https://doi.org/10.1007/978-1-4842-4324-4_8

an agreed-on process, development can become chaotic and end in failure. Luckily, your engineers know what they are talking about.

The Waterfall

The waterfall development model consists of distinct stages that are sequential. Let's say you are in the market for a new house. Congratulations! You cannot find anything that you want, so you decide to build your own.

First, you meet with an architect to define your timeline and requirements. These include details such as materials to be used and the number of bedrooms and bathrooms you want. After finalizing your requirements, you work with the architect to put together a blueprint. Next, you gather a construction team and break ground on the house. The general contractor tells you to come back in ten months when the entire house will be finished. When it's done, you visit the property and take a tour. In a way, you are testing it to make sure it was built to specification. Assuming there are no gaping holes in the walls or missing staircases, you will hopefully move into the house and invite me to your housewarming party. This development process followed the five stages of the waterfall model: requirements, design, implementation, verification, and maintenance.

Here, the inputs and outputs are well-defined, and backtracking to an earlier stage is incredibly expensive. If, for example, in the *implementation* phase of construction you wanted to change the layout of the house, imagine the costs of demolishing, redesigning, and then reimplementing the project. For this reason, the waterfall model recommends that every detail of the entire project is confirmed before moving on to the next stage.

A More Appropriate Approach

If you have been exposed to software development, you know that knowing every detail of the whole project is never possible. Who knows what new technology will be out in a few months? Maybe we will want to use that. What if the aspects of the code do not correctly work the way we expect them to? We might have to engineer it in a different way. What if we find out that our initial page layout is too confusing? All of these issues happen regularly, so a long-term and sequential process cannot necessarily be applied to software. Additionally, unlike home construction, there is room for more rapid iteration. Software typically does not require the same capital commitment as home building. Therefore, characteristics that support the waterfall model do not apply to software.

Iterative and Incremental Development

In the 1960s, an evolutionary model called *iterative and incremental development* (IID) emerged and was ultimately adopted by many software developers. It broke the long-term waterfall model into mini-projects, called iterations. For example, we can break the construction of MyAppoly into the development of its features, including transferring money, uploading photos, and exporting data to a spreadsheet. Together, these iterations form the whole project, but each iteration is a complete standalone aspect of the end goal. The export to spreadsheet feature should be independent (from an engineering standpoint) from the photo uploading function. Each incremental iteration adds something new to the existing product. "Tune-up" iterations enhance older features.

The many philosophies of development under the umbrella of iterative and incremental tend to recommend that an iteration be somewhere between one and six weeks. These iterations have deadlines, a practice called *timeboxing*. In 1955, Cyril Parkinson articulated his eponymous law, which states that "work expands so as to fill the time available for its completion."[2] By Parkinson's law, if you have one week to do something that you really know will only take two days, the entire iteration will still take one week because you will be unproductive. Timeboxing is designed to prevent such procrastination. Timeboxing also prevents scoped work from expanding. Say that a feature is timeboxed to a week, but it ends up being more technically complex. By limiting time spent on this feature to a week, we can reevaluate it at the end of the timebox and decide to finish it only if it is a priority.

Rapid iterations are needed to accommodate the unpredictability that defines software development. You can iterate on a single aspect multiple times, and so this development model is not as rigidly sequential as the waterfall model. If you were to apply this process to house construction, an iteration might consist of a single bedroom. Subsequent iterations could add different bedrooms or even redo a particular bedroom based on feedback from the prior iterations.

One concern with IID processes is prioritization. If someone breaks apart a large project into several iterations, which ones are tackled first? IID processes have two approaches to prioritization: risk-driven and client-driven. The risk-driven approach recommends working on the riskiest parts of the project first. These iterations could consist of the most integral parts of the application or the ones that are most complex and likely to fail. By building these first, you can be more confident that the project is feasible and will proceed as expected. The client-driven approach, as the name suggests, says that customer feedback

---

[2]Cyril Northcote Parkinson, "Parkinson's Law," *The Economist*, November 19, 1955. Available at http://www.economist.com/node/14116121.

and demand should influence prioritization. Such feedback-driven decision-making is considered *evolutionary* and *adaptive* in nature because it provides the flexibility valued by developers in place of the waterfall model's rigidity. These prioritization strategies are not mutually exclusive, but they are useful in framing the discussion within your own team.

There is a difference between *delivery* and *iteration*: not every iteration has to be delivered (made available for use by your customers). You might want several features to be built before releasing a new version, for example. With that being said, feedback from your deliveries might inform future iterations and deliveries, and so it's strategically advantageous to release iterations to customers periodically.

Agile Development

Today, many companies have shifted to an *agile* methodology. Agile development, like the IIDs already mentioned, is a philosophy of organization and management. Some say it is a type of IID, while others argue that it borrows from but significantly expands on IIDs. Whatever your standpoint, almost everyone, from the titans to the up-and-coming web startups, wants to say they have an *agile development process*.

Agile development displays many of the same qualities as IIDs. The product backlog consists of all the things yet to complete, organized according to some prioritization methodology. The teams work on a few iterations at a time, and iterations are released to users for feedback as soon as they are complete, creating more customer-developer collaboration than in most IIDs. This feedback adds and reprioritizes items in the product backlog. *Agile* does not refer to the enhanced speed of your team's process, but rather to the philosophy that describes the methodology of building it. It holds four core principles:

- Individuals and interactions over processes and tools
- Working software over comprehensive documentation
- Customer collaboration over contract negotiation
- Responding to change over following a plan

Agile development emphasizes team dynamics and the overall work environment. Rather than outline characteristics, I describe them from my own experience of agile culture at a startup. Agile development recommends group collaboration and open spaces and may favor an open floor plan over cubicles to spur creativity and remove rigidity. Whiteboards are plentiful so that self-managed teams can sketch out ideas, interact with their thoughts, and photograph them for future use. These brainstorming sessions are meant to be simple, as the agile development process encourages simplicity in all areas.

Instead of using complicated slide shows and spreadsheets to demonstrate a point, jot it down, take a photo, and send it around.

These companies are characterized by self-managed teams, where authority is shifted from the manager to the group. These teams brainstorm ideas for the next iteration, delegate their own tasks, tackle them in groups, reconvene to check on progress, and hold each other accountable. Older management beliefs dictated that people did not like work, avoided responsibility, and could only be motivated by negative incentives. Agile development fundamentally disagrees and has roots in more natural management theories, which propose that people feel empowered in teams, delight in tackling hard problems, and enjoy coming to work. The right environment will encourage the team to be more productive and to deliver a better product. Rather than dictate what must be done, project managers provide resources, maintain the vision, remove impediments, and promote principles that favor the team's adoption of agile culture.

Benefits

The benefits of agile development are clear. Rapid iterations allow teams to manage complexity and to refine their way of working together, thereby improving efficiency. The ability to incrementally add features and revisit old ones lowers the overall risk and delivers versions of the product more quickly to the customer. Customers love products that regularly improve, and companies can use the rapid and constant feedback to inform their product, which in turn improves the user experience. Your team will feel empowered by delivering changes frequently. Their buy-in leads to higher productivity, and in the end, both the business and the customers benefit. Because the development process adds productivity and quality improvements to the product, agile development is rising in popularity rapidly.

Release Management

How are iterations delivered to the customer? The process for releasing or *shipping* code to the public must be well thought out to minimize the risk of a buggy or faulty update. To *deliver* and *ship* code mean the same thing—that is, to make that code go "live" for the customers' use. There are three main environments or servers, as follows.

The first is the *development environment,* where all the coding and testing takes place. A new feature exists on its own server and cannot be accessed by the public. Therefore, any mistakes are localized to the company, at worst only slowing down the development process. Code in the development environment does not affect the "live" version of the site.

When the iteration is ready to ship (in the form of an update or new version), the new code must be uploaded to the *staging environment,* a different server that has an exact copy of the live application. Not all products use a staging server, but testing in the staging environment serves as another check to confirm that the new feature is ready for final delivery.

If the code seems to work perfectly in the staging environment, the update is ready for *deployment.* It gets moved to the production environment, which serves the public site and responds to all the site visitors. The release process is designed to ensure this server is only updated with clean code that works correctly. Of course, no process is perfect, and sometimes bugs creep in.

Conclusion

After discussing with your team, you decide to implement one of the development models discussed in this chapter. You also work with your engineers to develop an effective release management system, implementing several checks for each step of the release. Before introducing new features to customers, however, it is prudent to make sure they are bug-free and not going to create more chaos than they solve. Your next task is to learn how your team does just that.

Software Development: Debugging and Testing

In 2009, Toyota faced allegations that some of its cars had sticky accelerator pedals and, in January 2010, recalled 4.1 million vehicles for repair. The defect was a *bug* or flaw in the company's product. We can draw an analogy between Toyota's product and that of any Internet company. Facebook, Google, and MyAppoly all have and will continue to have bugs, but great teams identify them in the development process before the product is released for customer use.

It is unrealistic to expect that every line of code your team produces will work correctly the first time. Just as proofreading and editing written materials before publication is a critical part of the writing process, *debugging*, the process of removing errors, is an important step in the process of writing code. These errors include not only improper spelling and other syntax-related problems but also problems in the structure of the code or how code interacts with other code. Because your team and users will discover

© Vinay Trivedi 2019
V. Trivedi, *How to Speak Tech*, https://doi.org/10.1007/978-1-4842-4324-4_9

bugs in the product regularly, debugging is a continuous process that takes an unpredictable amount of time. As MyAppoly grows larger and involves a growing number of engineers, quickly identifying the source of a bug will become increasingly difficult. Understanding how your engineers approach debugging at a conceptual level will help you see how experimental and variable the process is.

A Bug's Life

Simply put, a *bug* is an error. The unusual term has an interesting history. Back when computers were coded using punch cards, Grace Hopper, a famous computer scientist, noticed some unusual behavior. Upon investigation, she found a dead moth interfering with one of the punch cards—computer science's first bug!

Today, bugs are (unintentionally) created by programmers instead of by physical bugs, and not all bugs are simple to fix. There are two kinds of bugs: syntax bugs and semantic bugs.

- *Syntax bugs* result from breaking the rules of the programming language. Your programmer might have forgotten a semicolon, misspelled a variable name, or tried to use a structure not provided by the programming language. These bugs are often easy to fix because a compiler, interpreter, or another tool can find them. If code were a Microsoft Word document, these errors could be caught by a spelling and grammar check.

- *Semantic bugs* consist of two types. First, *runtime errors* cause a program to crash and stop its execution. An example would be if the program tried to open a file that didn't exist. Second, *logical errors*, which do not disrupt the completion of a program, produce unexpected results, such as adding something instead of subtracting it. If you contradict yourself in an essay, you commit an error in logic, but as long as the grammar and spelling are correct, spellcheck will not flag it. Therefore, semantic errors are more difficult to catch, isolate, and resolve than are syntax errors.

Although programmers create all bugs, they are not all due to negligence. Some bugs emerge because code is updated that is no longer compatible with prior logic. Some bugs emerge because two engineers merge their code, and the two new features do not work perfectly together. Just because a bug exists does not mean someone has to be blamed.

The Debugging Process

The six-step debugging process laid out in this section is a distillation of the "13 Golden Rules of Debugging" of Grötker and his Synopsys colleagues.[1] These six steps are not canonical, but you may be confident when engineers on your team say they are "debugging" an issue that they are following some variant of these steps.

1. Track the Problem

To manage the errors of MyAppoly effectively, keeping track of all the bugs is a great start. Maintaining proper documentation on every bug you encounter can help you resolve future issues more rapidly. Additionally, tracking bugs can help your entire team stay in sync about problems and their solutions. Many out-of-the-box tracking systems operate as follows. When a bug is found, an issue or "ticket" is created that describes the bug. Its status remains *open* until it is resolved, in which case the status is switched to *closed*. Several intermediary statuses exist, such as *in progress*, and tickets can be assigned to different team members as a delegation method.

What constitutes proper documentation? Because any program input can influence bugs, everything that might have contributed to the problem should be recorded. This includes things like the particular action the user took, the operating system, browser, actual behavior, expected behavior, and pretty much anything else you can think of that might be relevant. These variables will either be *static*—properties related to configuration and compatibility that don't change, such as the operating system and browser—or *dynamic*—referring to aspects of memory and network that change frequently, such as how many other programs were running at the same time.

2. Reproduce the Problem

After you learn of a bug and record its details, the next step is to reproduce the problem. This involves re-creating the environment, as well as checking whether the behavior matches the one reported. This process is somewhat manual at first, but you want to re-create the program failure so that you can use the same test to confirm whether you have properly fixed it.

---

[1] Ibid.

3. Run Tests

After every adjustment to the code, try to reproduce the failure to see if the last update eliminated the problem. Manually reproducing the problem will be tiring, so a better way to automate the tests will allow you to identify the problem. There are several techniques for doing so; the availability of each one is based on which programming language you are using, the specifications of the system overall, and what tools you are willing to buy.

The scientific method will inform your debugging approach: hypothesize, edit code accordingly, test, evaluate, and repeat as necessary. After testing the program following every code adjustment, log the change to create an audit trail so you can backtrack if you need to restore the code to its original state. Additionally, it is good practice to create different versions of the file for every change you make.

How will you test the functionality of MyAppoly before you launch the new version? You can't waste any time because you want to replace the buggy code released to your users as soon as possible. The tests you run will typically occur at one of the three levels: the presentation layer, the functionality layer, or the unit layer.

Although you can test at any of the layers, you will typically choose one based on ease of execution, ease of interaction, and ease of result assessment, among other considerations. Let's dive into this quickly.

The *presentation layer* is the view component in the model-view-controller description of our application. To test from the presentation layer, you interact with the application as a user ordinarily would, using the mouse and keyboard. Let's say there was a bug in checking some account value on MyAppoly. After a $100 deposit, the account value still reads $0. You could use the presentation layer to check if you fixed the bug by refreshing the web page and reviewing the account value. Testing at this layer is generally feasible and quick. Even so, visibility is limited because you cannot necessarily see how the back end is making the calculation just by looking at the final value.

The *functionality layer* consists of all the code that defines the functionality of the program—essentially the controller and model parts of MyAppoly. To test at this layer, programmers typically write a smaller program that interacts with the code in the larger application. For example, a programmer may want to write a small program that creates a fake account and adds $100 to it, checking that the account balance after the deposit is actually $100. Running these functionality tests is typically as easy as clicking the refresh button on a browser, and automation allows testing at scale. Creating these tests may seem easy, but there are some startup costs for developers to write each one. If you expect to be running the same test a lot, however, it might be worth the investment.

The last layer is the *unit layer*. If you could break down the full program into small, one-operation chunks, you would have a unit. In the account example, each of the following could represent a unit: retrieving the old account value, getting the deposit value, calculating the new account value, and storing the new value. Testing these specific units can be automated with a script. These tests give the programmer the most detailed look at how the program is operating at the building block level. Often these tests are of equality, meaning they will test if the expected output equals the actual output. If it is true, the tests will continue; otherwise an error will be reported. In this way, the bug can be immediately traced to a particular area of the program. In the current example, there are several units. First, you can check the value by which the account is going to change. If you entered $100, let's make sure the back end is actually interpreting it as $100. Now, let's get the existing value of our account. Are you getting the right account? Finally, you can add $100 to the existing account value to obtain the final value. Is the final value correct? By breaking it down into a step-by-step analysis, you can determine the specific step at which the failure occurs.

Your programmers have various tools and techniques to aid their efforts to track and test bugs. An easy debugging methodology involves the print function, which displays anything you tell it to. In the example of updating your MyAppoly account value, you could add print statements in between every line of code. You could print the old account value, the amount you want to change it by, and the new account value. If any one of those numbers were off, you can then see exactly where things went wrong. In this way, print functions can be used to create unit tests. But adding print statements everywhere in your code, as you can imagine, is messy and tedious, as you'd have to read through a lot of output to find each bug. Luckily, there are better ways.

One tool you can use is a *debugger*. Debuggers essentially stop a program from executing so programmers can make observations. By setting *breakpoints*, a programmer can stop the code in place and see the values of all the relevant variables. Debuggers also allow programs to *trace* through code, pausing after each line, so that the programmer can determine which line will execute next and how the variables should change. Presumably, if an error occurs, the programmer will be able to identify where it occurred, and by tracing through the logic of the code with the computer, programmers can more easily spot the bug. An example of a debugger you might hear of is the GNU debugger GDB, which is run from the command line and is typically associated with the C programming language.

4. Interpret Test Results to Identify Bug Origin

With every test, you come closer to understanding the root cause of the bug. If you ask your programmers, there is nothing more fulfilling than finding the logical error or missing statement that caused so much pain. Keep in mind that the bug could exist pretty much anywhere, whether in your source code, in the compiler (if you are using a compiled language), or in a third-party library or API.

5. Fix the Bug Locally

Now that you've found the bug, it's time to fix it. Even if the fix requires you to reprogram an entire feature, you know *why* the program is crashing. After you fix the bug, your team will run regression tests, which test if all of the other functionalities of the site are still working. Fixing one bug and introducing another is hardly the right solution. You may also write additional tests to ensure such a bug doesn't slip by unnoticed next time. Assuming your regression tests pass, you can feel confident that you have eliminated the threat without creating more chaos.

6. Deliver Fixed Product

Once you have fixed the bug and run confirmation tests, the MyAppoly team will update the public version of the site so users are no longer accessing a buggy application. This follows the release management process outlined in Chapter 8.

Conclusion

Now you understand a bit more about the process and techniques your team will use to make MyAppoly achieve perfection. With these tools in hand, your team has built, tested, debugged, and launched a working version of MyAppoly. Improving the product is always a goal, but the one taken up in the next chapter is mission-critical: getting users.

Attracting and Understanding Your Users

Users can learn about your website through many different ways, each called a *channel*. One might be through *social media*. For example, new users might learn about and join MyAppoly via a Facebook post or a tweet. Another channel could be through *referrals* (jargon for word of mouth). Yet another could be through an email marketing campaign. As you can see, there are several ways to reach potential customers and convince them that it is worth their time to consider your website. This chapter focuses on the rise of search engines and the two prominent channels associated with them: organic and paid search. *Organic search* refers to the results that pop up from standard search. When you search for something online, you notice paid advertisements relevant to your search terms at the top of the *search engine results page* (SERP). *Paid search* refers to the presence of these ads.

Businesses collect data on their customers to understand how they are performing. Grocery stores, for example, price inventory, stock items, and change their layout based on information like who their customers are, which products they purchase, and which aisles tend to be most popular. To learn some of this information, grocery stores intelligently introduced loyalty cards.

Customers see this card merely as a tool to obtain discounts, but in reality, the customers are empowering the stores with their personal data.

Loyalty cards work for grocery stores because customers who enter the store rarely leave without purchasing anything. Online customers are different. If you were only able to collect data about your MyAppoly customers when they made purchases, you would lose a lot of valuable data. Just like a grocery store, you want to figure out who your customers are, what they want, and what they think about your product. Since you cannot feasibly survey every user the way you could customers of a grocery store, you need to find an alternate source for the information.

This chapter describes how search engines can be leveraged to attract users and how websites track what those users are doing. We also briefly discuss the new data protection regulations designed to protect our personal information from being leveraged without our explicit permission.

Search Engine Optimization

Today, we take search engines for granted, but the ability to search across the Internet was nonexistent until 1990, when a search engine named Archie allowed users to search by filename. From this point, several players worked on creating better search engines. In 1994, Yahoo! created a database of all web pages it could find manually and gave users a way to search through the catalog. Unfortunately, their approach was limited in effectiveness because the manual process of collecting pages was sluggish. AskJeeves proposed a new method of asking questions that was not exactly a true search but enabled knowledge sharing. These early players fought over their respective strategies. Little did they know that they were going to lose the fight to two Stanford University graduate students named Larry Page and Sergey Brin. In 1998, these two students worked on a way to scan the entire Internet and store a description about each page to recall and display websites based on a user's search query. The number googol (a one followed by one hundred zeros) inspired the name of their search engine. Google was born, and the nature of how we would use the Internet fundamentally changed.

Modern search engines store information about all websites and associated keywords. When you search for a term or phrase, they match your search to the descriptive keywords linked to each website.

How do modern search engines actually collect the name and keywords of websites? Gone are the days of manual entry into a catalog! Necessity is the mother of innovation, as the adage goes, so as the number of web pages grew at an ever-increasing rate, the manual process of logging them became untenable. Smart Google engineers refined their ability to "crawl" over the Internet using *bots* called *spiders* that scan a web page and follow any links contained on the page. If they follow every link, they'll pretty much discover the entire Web. As they scan each page, they record keywords that appear, the number of pages that link to it, and several other characteristics that help to describe the page and to assign it a rank comparing its relative importance to others on the same topic. The rank determines the order in which it is displayed when someone searches for a topic on Google, which is valuable since most users don't look past the first few pages of search results; a higher ranking drives more traffic to your site. The process of searching for a topic and clicking a ranked article is organic search, and *search engine optimization* (SEO) refers to the process of improving your rank on Google and other search engines to increase the likelihood of acquiring users through organic search.

In order to understand search engine optimization, you must also learn how the actual matching of a search query to database keywords works. Because queries can vary based on capitalization (e.g., *Tennis* vs. *tennis*), plurality (e.g., *rackets* vs. *racket*), spelling (*racket* vs. *racquet*), and the inclusion of nonessential words and phrases (e.g., *the* and *a*), search engines typically remove all these variants before checking the query against the database. After the query is matched and all the corresponding web page results are selected, Google must rank them before presenting them to the searcher. The best results always seem to be at the top, thus minimizing the time a user spends finding the right link. In some ways, this is where the real magic of search engines lies. The actual crawling of the Internet and creation of the database are predictable, but the information collected and how that information is used to rank and display results are the differentiation factors in search.

Google's ranking algorithm is called PageRank, and it has become famous over the years. Though the exact algorithm is not public, we have an idea of what factors it considers, and, as a result, you may be able to use Google's ranking methodology to inform how you build your web pages.

One thing the algorithm targets is keyword density. Pages that reference a keyword frequently are more likely to be relevant than ones that mention it less often. A page that mentions "tennis racket" one hundred times is more likely related to tennis than a page that contains only one occurrence of "tennis racket." Search engineers quickly realized that web developers might clutter the page with redundant occurrences of the keyword in order to obtain a higher rank. Smart, huh? Google and other search engines are smart too, so they monitor such behavior and flag it for rank demotion.

Search engines also look for keyword proximity and prominence. They want to make sure that the words *tennis* and *racket* appear next to each other because if they do, the chances that they relate to the search "tennis racket" are high. Keyword prominence can also indicate the relative importance of the article. Is *tennis racket* the title of a web page or merely in the footnote of an image? This also factors into the rank, as most would agree it should.

Last and certainly not least is link popularity. If a lot of other websites have links that lead to your page, the chances are good that your page is high in importance. This is especially true if important websites (also determined by link popularity and the other factors) have links to your page. To continue the example, if the US Open tennis website linked to your page on tennis rackets, that's a critical piece of information and Google gives you credit for it. What they don't give you credit for are links you put on your own page that take you to other well-known sites. In other words, including a link to the US Open site on your own page will not affect your rank. Anyone can put links on their website, so that's not special.

Based on this knowledge, you can optimize your web pages to achieve a higher rank through the process of SEO. You have direct control over things that appear on the web page. A clear, descriptive title; strategic use of keywords (keeping in mind proximity, density, and prominence); and descriptions that you can add to the meta tags (you can use these to provide data to the Google bots) and image tags in HTML all influence your rank. Regularly updating the content on your page might also be perceived as a sign of credibility, as are search-specific site maps that define and describe for crawlers your page layout, description, importance of each page, and so forth.

As you may have guessed, it is a lot more difficult to control your link popularity. Short of forcing others to include links to your web page, there is little you can do other than building your credibility and awareness.

The better you understand how the ranking algorithm of search engines work, the better you will be at SEO.

Search Engine Marketing

By advertising your company on Google or Bing, your website would be featured on some of the most highly trafficked real estate on the Internet. This section will discuss the principles behind how this works.

The concept of *search engine marketing* (SEM) is quite simple. Let's continue to suppose MyAppoly is a tennis website. You know that users searching for tennis rackets are likely in the market for new tennis gear. In order for your ad to be displayed when a user searches "tennis bag," you must "purchase" the keyword "tennis bag." When a user searches for the term "tennis bag," your advertisement will pop up on the side, and, if the user clicks your link, you will be charged the amount at the rate you purchased the keyword for if the user clicks your link. There is an additional layer of complexity to this, however.

Many people want to buy the same keyword. All tennis websites probably want to buy the keyword "tennis bag." How does Google know which ads to show in the space provided (again, at the top of the results page)? This is determined by a real-time auction based on two factors: the bid and the quality.

When you create your campaign and ads, you specify how much you want to bid. Naturally the higher you bid, the more likely it is your ad is shown over other ads from competitors who purchased the same keyword. This makes sense since Google wants to maximize its revenue and, all else equal, a higher bid is better than a lower one. Quality is equally important because Google is paid only when a user clicks the ad. Let's illustrate by example. What if MyAppoly started buying ads for lacrosse helmets thinking that any athlete looking to buy equipment may also be interested in tennis gear? MyAppoly may win the auction for lacrosse helmets, but as soon as users click MyAppoly's ads, they will realize that MyAppoly is a tennis website and not related to lacrosse at all. They will probably hit the back button immediately. Google can track this activity, and if they notice that your ads are not serving the customers what they actually want, your quality score will go down. Therefore, if you want to win auctions, you need to have well-targeted ads (good quality) and bid a value that is likely to be higher than your competitors' bids.

This is known as a *performance-based advertising model*, which is nice because, unlike traditional advertising methods, you only pay if users click your ads. The incentives for all parties are aligned. The advertiser only pays if the ad is clicked. Google collects money by showing ads that users are likely to click. The user will only see ads that are relevant. Compare this model to that of yellow pages, in which advertisers pay a flat amount for placement regardless of the number of people that see it and users have to sift through hundreds of irrelevant ads. This is inefficient in today's world with so much access to highly targeted data.

Search advertising is effective because of the context the search query provides. Wasting money on advertising to the wrong population is a danger of the past because you only purchase keywords that describe your target market. Additionally, when a user clicks an ad, you can control where they go on your web page. Therefore, rather than directing all users to the homepage of MyAppoly, why not direct people who click the "tennis racket" ad to the list of tennis rackets on MyAppoly?

Analytics

Analytics refers to software programs that generate metrics describing how users are using your site. The history of analytics begins not as a set of user tracking metrics but rather as an error log that stored user actions to assist debugging.[1] Early developers created automatically generated web documents that logged errors, hence the name *web logs*. Over time, however, business

---

[1] Avinash Kaushik, *Web Analytics 2.0: The Art of Online Accountability and Science of Customer Centricity*. Sybex, 2009.

and marketing employees realized that the web log data had value beyond debugging. This data could be used to understand the customer.

In 1995, Stephen Turner wrote a program called Analog that analyzed these convoluted log files, extracting the portions of it that were useful for analysis. Webtrends, a later program, helped people visualize the data stored in these logs. The goal of collecting and visualizing meaningful data about website usage is motivated by a desire to make decisions based on information about users and their habits.

What are these data that web application companies are interested in collecting? There are several types, but the one discussed here is *clickstream data*. When users visit MyAppoly, they leave behind a trail made of all the pages they visited and what they clicked. This record of a user's activity defines the term *clickstream data*, and there are four main ways of collecting it: web log, web beacon, JavaScript tagging, and packet sniffing.

Web Log

When a user visits a website, the browser sends a request to the server where the website lives. The server updates its logs with something like "User X visited page at 12:00 pm" before sending the page back. That's a *web log*. One downside is that any server request is logged, even the bots of search engines. If you want to monitor only human activity, this may not be the best method to use. Additionally, unique visitors are difficult to identify which can complicate a marketing analysis. You do not want to count the same person twice, or else your numbers will be off. Lastly, do you recall the term *cache*? It takes time for browsers to send requests to servers and to receive the page with all of its images and other resources. Therefore, browsers save a copy of the website locally to minimize the server interaction for efficiency purposes. Web logs will not kick in when the server is not utilized, so all activity on cached versions of this web page will not be logged. This is an example of server-side data collection (the software used to store the data lives on the server).

Web Beacon

You might be familiar with web beacons if you have received an email with images temporarily hidden until you authorize the email client to show it. Why? Web beacons typically are 1x1 pixel images that are transparent and live on separate servers from those holding the web page. To load an image, recall that browsers have to issue additional server requests. When browsers request this web beacon from a third-party server, their activity is logged. The use of third-party servers offers a bit more flexibility over what you want to collect (either include the web beacon or do not) and allows you to aggregate data across multiple web pages—that is, you can issue requests for this web beacon from many websites, all of which must go to the same server to get the 1x1 pixel image. As a result, all the data are stored on the same server, even though the requests came from different sources. With that being said, since this method relies on an image request, any tool that blocks image requests prevents you from gathering useful data. Spammers use this to verify that your email account is still active, so some email clients disable images until you authorize their authenticity. Some services also use web beacons to enable read receipts.

JavaScript Tags

JavaScript tagging changed the nature of web analytics. If you recall from our discussion on JavaScript, the language has listeners that wait for particular events (such as the clicking of a button) before executing certain code. In the same way a JavaScript listener might wait to fire code showing a pop up until a user has clicked a button, JavaScript could use the "page has finished loading" event as a trigger to execute code. If you could make that code collect information about the user, the time, and a description of the web page, you could use JavaScript as an analytics tool. This is exactly what has happened. As long as you include the line of JavaScript code that fires when the page loads, you are good to go. Even if your page is cached, the JavaScript code will execute when the user views it, so it beats web logs in that way. You can also add special tags to the JavaScript code that help you filter data more easily afterward.

There is slightly more action than this simple explanation because cookies can play a large role. When you visit a website, it might put something called a cookie on your browser that stores information about you, what you clicked, and other relevant information. Next time you visit the website, it will look for that website's cookies, and if it finds one, it knows that you have been to the site before and can use the contained information to customize the page for you. If the cookie stored information about the shoes you were shopping for, those shoes might appear at the top of the page the next time you visit. Back to the analytics: JavaScript code uses cookies to log activity. It stores a cookie on your browser, and after other parts of the code are finished executing, the information then gets passed to a server for storing. These servers are owned either by the company or by third-party analytics providers.

Whereas before analytics were entirely managed by a company's internal IT team, whole companies have now emerged with the sole purpose of helping others with analytics. These companies provide some JavaScript code for others to put on their websites, and they take care of the rest. The value they provide is also in the form of visualization tools to help you understand your data. Google provides analytics services for free through Google Analytics, which works by storing data in a log file. Every few hours, Google processes these log files and then makes the data available for the user to review on its Analytics website.

The benefits of this approach are many, the primary one being ease of use. Simply by pasting code the analytics company provides, you have a full tracking and visualization system up and running. Voilà!

With that being said, some users are starting to turn off JavaScript, so you can expect a small amount of data leakage with this approach. Also, do keep in mind that if you use a third-party analytics provider (i.e., you do not build and collect all of this information yourself), your sensitive MyAppoly data will be in the hands of other people.

Packet Sniffing

Packet sniffing essentially creates a middleman between a website visitor and the MyAppoly server hosting the website. After a user requests the MyAppoly homepage, it passes through a mix of software and hardware called a *packet sniffer* where information about the user is recorded before being routed to the server. When the server responds with the web page, the files go back through the packet sniffer before getting routed to the user. That's it. On one pass, it gets the user data, and on the other pass, it gets the web page information.

What is convenient about this approach is that no additional code needs to be entered in the actual source code and it can easily intercept all requests made to your server, thus covering your entire website's needs. The drawbacks are the actual setup process, which might be convoluted, and requests to cached versions of the site will not be recorded. You also have to be careful with packet sniffers since they will intercept a lot of sensitive information such as users' passwords and credit card numbers. Dealing with this data in a privacy-respecting way is crucial.

Visualization

The particular way data are displayed influences the insights you can draw. As a result, many players are trying to make this process easier still. The next stage of analytics, therefore, seems to be not so much aggregation-focused but visualization- and processing-focused.

Click density (aka *site overlay*) provides insights on what users might be thinking based on what they are clicking. The eventual development of *heat maps* accomplishes the same thing. Heat maps are images of a web page colored to indicate where users are looking based on the mouse movements. Dark red is used for areas in which the mouse often hovers, whereas blue areas are locations on the page where the mouse rarely goes.

Analytics should seek to answer not only the *what* (how many views, how many signups, and so forth) but also the *why*. The focus has been redirected to tasks, with funnels being the popular way to visualize task completion. In this way, companies can see exactly how many people drop off at each stage of the process.

The list of visualization types could continue indefinitely, but the main point is that these data are valuable to companies, and they are looking for any way to make sense of them. Presentation matters.

GDPR

With all of this talk about tracking, it is important for the MyAppoly team to be aware of the serious concerns around privacy. As of 2016, we have introduced significant changes to our existing data privacy regulation, most notably General Data Protection Regulation (GDPR) which attempts to standardize data protection law across the EU countries with a more comprehensive and stronger solution than we had previously. Enforceable starting May 2018, GDPR requires businesses to disclose any data collection before it does so, including the data's purpose, storage time, and sharing plans. The business may only move forward in processing personal data if it receives explicit consent from the user, which can also be revoked at any time. GDPR is far more comprehensive than we will get into here, but this should be central to the conversations your team has on data tracking, collection, storage, and processing.

Conclusion

Keep in mind that several tools have been omitted in this discussion. If you sign up for Google Analytics, you will notice that there are many helpful methods to segment the data and extract hidden insights. For example, Google Analytics allows you to see what words people search on search engines before coming to your website. This is called *site search analytics* (SSA) and provides context behind a user's actions, unlike the data produced by clickstream analyses and heat maps, which require inference.

Now that you have been primed on the basics of search engines, tracking, and data privacy regulation, MyAppoly is ready for the big leagues. It's ready to grow, but can you handle the growth?

Performance and Scalability

As an eager entrepreneur, you monitor MyAppoly's usage carefully and address any issues immediately. You spend time sifting through the analytics reports from your team, and you notice signs of success. The number of users you have is snowballing. Soon, your team approaches you and asks to discuss how to improve the site now that it is attracting unprecedented traffic. Some engineers note that certain pages are much slower than their potential and that the current setup cannot support many more users, especially at the rate of growth the metrics suggest. For example, each server can only handle a certain number of requests per second, so the website currently takes too long to load. These are good problems to have, but they still need to be addressed.

Performance relates to a website's speed: how fast it loads, how many bytes are transferred, and any other metric that measures the rate at which a user receives requested information.

Scalability refers to the ability of a web application to adjust to a growing amount of work. It is about how well your site can accommodate more users, more data, and more computation—not just one more or a few more, but thousands or even millions more.

© Vinay Trivedi 2019
V. Trivedi, *How to Speak Tech*, https://doi.org/10.1007/978-1-4842-4324-4_11

You can see that the two are not the same thing but might arise together. If your user base is growing, you want to make sure your site is operating as efficiently as possible. Your team prepares a primer on techniques they hope to implement. The list is not exhaustive but gives you a sense for how companies approach these issues.

Practices to Improve Performance

The back end and front end can be optimized to improve performance. However, the back end accounts for only a fraction of the time it takes for a page to load. Most of the time is spent downloading components for the user interface—such as issuing HTTP requests for the assortment of elements that appear on a given web page. This might influence how you prioritize and allocate resources, though page load time is just one aspect of performance.

Back-End Considerations

Applying the good design principles already mentioned can create efficiency gains. For example, abstraction allows you to reuse code effectively in different sections of your application.

Additionally, many successful technology companies rewrite their code to make things more efficient. Optimizing database queries and other operations (such as algorithms for determining recommendations for movies on Netflix) is one step. Another is to rewrite the application in a different programming language completely. Recall that lower-level languages provide more efficiency (Chapter 3). As a result, many technology companies rewrite their application to utilize these languages. Twitter, for example, transitioned from Ruby on Rails to Scala and Java. Facebook transforms its PHP code into highly optimized C++ code programmatically. The list continues, but because migrating code to a different language is an intensive process, companies need to consider thoughtfully when their current programming language is no longer the right tool for the job.

Front-End Considerations

When a user visits your website, only a small amount of time is spent on the initial HTTP request for the HTML document. A bulk of the time is spent downloading all the various resources associated with the particular web page. This section highlights some practices, tools, and methods for all front-end developers to think about when performance is the objective.

Fewer HTTP Requests

Reducing the number of HTTP requests required to download page resources shortens page load time. There are a variety of methods for reducing the number of requests, such as combining scripts and stylesheets. We'll discuss two image-specific approaches: using *CSS sprites* and *image maps*.

Imagine that you want to show the first half of your logo on the upper-left corner of your site and the right half of your logo on the upper-right corner. One approach would involve designing two elements representing each half that would require two requests to display. Instead of this approach, we could issue only one request for the full logo, and use different CSS attributes to hide portions of the image we do not want to display. This is a *CSS sprite,* and it helps reduce the number of requests we might need.

Now say that you want to include an image of your team on the site, where clicking any of their faces would redirect the user to their LinkedIn profile. Doing this with individual photos would require the browser to issue a separate HTTP request for each image. However, by using a group photo, you can use an *image map* to associate different links to different areas of the image, reducing the number of image HTTP requests to one.

Cache

If your users return to the same page repeatedly, it makes little sense for them to re-request all of the page's resources each time. You don't ask your best friends for their cellphone numbers every time you meet, do you? At some point, they would just ask you to save it. In the same way, your browser gets fed up with asking the server and instead stores web pages in a *cache.* Examples of cached resources include images, CSS files, and JavaScript files, all of which change relatively few times compared to the HTML; you might want to add a new paragraph of information to your site, but you're unlikely to change your logo or page styling frequently. Placing CSS and JavaScript in external sheets enables their caching.

Compression

If you've ever downloaded a large file, you may have noticed it downloaded as a smaller .zip file that had to be unpackaged. To create this file, we use a process called *zipping.* By reducing the size of the response, the time it takes to transfer them from the server to the browser is reduced. Keep in mind that there are costs entailed with compression. It takes additional processing power to compress files on the server and for the browser to decompress the files. Therefore, it does not make sense to compress every possible resource.

Another compression tool is *minification*, the process of deleting all the unneeded whitespace, comments, and characters from a document. When you are writing software, indentation and comments make your code readable. However, when transmitting code from the server to the browser, they can add a lot of unnecessary bulk.

Placement of Scripts and Stylesheets

Remember that stylesheets (CSS documents) and scripts (JavaScript documents) are often externally linked to an HTML document. Where you link these documents will determine how the page will load. Let's say it will take a page five seconds to load. In the first scenario, the screen is blank for the full five seconds, after which the entire page is instantaneously shown. In the second scenario, the page loads in parts over the five seconds until it is finished. Which is better? Many would say the latter because users are uncomfortable with unresponsiveness and like to know that something is happening. To achieve this effect, known as *progressive loading*, you should link the stylesheets at the top and the scripts at the bottom of the HTML document. Since many browsers do not load elements until they have processed the stylesheet to avoid redrawing elements, it's faster to have the stylesheet at the top. Scripts should be placed at the end because content below the script is halted from rendering until the script has been fully downloaded.

Infrastructure Changes

Most web application endeavors, such as MyAppoly, start with a single server linked to a single database with a set amount of storage. As more users start using your application, you need to add servers to respond to all of their requests. After more users join, however, and they begin interacting with each other, writing all of the data you collect to the database proves immeasurably slow. To interact with the database quickly, we *replicate* our databases, or create multiple copies we can interact with.

The process of replicating databases is tricky. If the queries are only "reads"—meaning the user wants to access information stored in the database—multiple copies of the database would work to satisfy a growing user base. What if a user wants to "write" or change what is stored in the database—say, add a friend? You have to make sure all the databases get updated. It's probably inefficient to add your "new friend" to all copies of the database, so you resolve this situation with a primary database designated as the write database, such that all updates have to be handled by this database alone. All other replica copies of the database are used purely for reads. In this way, databases can be scalable.

After adding more servers and databases, how do you actually direct users to the different copies? You use a *load balancer* to send requests across available servers in a way that spreads out the requests. Load balancers can employ any number of sophisticated algorithms to decide what the optimal assignment pattern is. The simplest is the round-robin approach, which directs incoming users to servers in a fixed pattern. After sending a user to Server A, the next user is sent to Server B, the next to Server C, and so forth. However, this method may be suboptimal because it does not consider the actual availability of the servers. You might accidentally assign two heavy users to the same server, even though other servers are being used more lightly. An assignment based on usage could work better. Load balancers provide reliability to your system, as they can redirect traffic if a server goes down.

Conclusion

You come away realizing that performance is key, and you now have a perspective on how developers think about improving the load time. You realize you will have to pay for more resources to support more users as your web app starts to scale. Additional resources require a trade-off between cost and performance—a good problem to have. As you grow, you realize you have a blind spot: security. The next chapter exposes you to some of the big attacks and the methods for stopping them.

Security

Special thanks to Ashi Agrawal for contributing this chapter.

In late July of 2017, credit reporting powerhouse Equifax disclosed that a group of unknown hackers had compromised the personal information—including driver's license numbers and Social Security numbers—of over 145 million customers.[1] The attack was the latest in an ever-growing wave of public breaches and hacks that had already struck Target, Ashley Madison, and Citigroup, among others.

While the Equifax breach was especially shocking due to the data's sensitivities, such an attack could strike *any* company. Hackers—people who look for cybersecurity flaws—may target an organization's digital assets for a variety of reasons, even to enact personal revenge or to claim glory. Some want to sell data to the highest bidder, while others relish the thrill and the challenge. Not all hackers have ill intentions; ethical hackers (also called white hat hackers) look for these flaws so that companies can fix them. Even so, the digital world is rife with nefarious attacks.

Regardless of what your application does, an attack can cause you to lose valuable computing power and data or lead to a business interruption. Even if a data breach doesn't expose consumers to identity theft, it can still reveal

---

[1]Alex Johnson, "Equifax breaks down just how bad last year's data breach was," NBC News, May 8, 2018. Available at https://www.nbcnews.com/news/us-news/equifax-breaks-down-just-how-bad-last-year-s-data-n872496.

©Vinay Trivedi 2019
V. Trivedi, *How to Speak Tech*, https://doi.org/10.1007/978-1-4842-4324-4_12

sensitive data and cause consumer trust to evaporate. Beyond business reasons for enacting security precautions, organizations have ethical and legal requirements to remain secure. Cybersecurity should be of utmost importance for you and your company, and this chapter will discuss methods to keep yourself and your users safe.

What Is Cybersecurity?

Cybersecurity falls under the umbrella of information security, the practice of protecting electronic and physical data.[2] As the *Oxford English Dictionary* defines it, cybersecurity focuses on protecting electronic data, but this definition is limiting as it disregards securing computer resources that aren't data.[3] The field also deals with digital attacks that are not data-driven: consider hacks that dismantle power grids. These attacks don't compromise data, but they do use digital tools to wreak havoc. Therefore, this text defines cybersecurity as the study and practice of protecting electronic resources, such as data and computing power, from unauthorized parties. This chapter uses the term security in place of cybersecurity to reflect how professionals in this space use the terms.

Certain cybersecurity techniques have roots in cryptography, the study of securely communicating in the presence of malicious third parties. As a field, cryptography has evolved over the centuries from simple techniques— reversing letters in a message or using the children's language game "Pig Latin"—to complex, mathematics-based systems. These systems have since been crucial to some cybersecurity techniques that we'll delve into later.

After the rise of the Internet, cryptography and security quickly proved necessary in the digital space. The first computer virus is widely acknowledged to be the 1988 Morris worm.[4] The program took advantage of vulnerabilities in various network protocols to spread rapidly and embed itself on machines. While Morris initially intended the program to measure the size of the Internet harmlessly, it ended up devastating thousands of computers by consuming and overwhelming their processing power. The resulting computer failures caused hundreds of thousands of dollars in damages. Since then, attacks have

---

[2]Patricia de Saracho, "Cybersecurity, Information Security, Network Security, Information Assurance: What's the Difference?" Security Magazine, September 6, 2018. Available at https://www.securitymagazine.com/blogs/14-security-blog/post/89383-cybersecurity-information-security-network-security-information-assurance-whats-the-difference.

[3]Full definition from *Oxford English Dictionary* for cybersecurity: "The state of being protected against the criminal or unauthorized use of electronic data, or the measures taken to achieve this."

[4]Charles Schmidt and Tom Darby, "What the Internet Worm did to systems," The Morris Internet Worm. Available at https://snowplow.org/tom/worm/what.html.

become more sophisticated, and, with more users and devices than ever on the Internet, the possible impact has ballooned in size.

The CIA Triad

The CIA Triad provides a framework for understanding the attack landscape. This classic model of information security outlines three primary security concerns: Confidentiality, Integrity, and Availability. Imagine the following situation: Alice wishes to pass Bob a piece of information and to protect it from their adversary, Charlie[5]. Let's consider three ways Alice and Bob can feel secure about their transaction.

- *Confidentiality*: Since the message is a secret, we want it to stay secret. It is only ever visible to Alice and Bob, not to any third parties like Charlie.

- *Integrity*: The content of the message matters, so we want to make sure that Charlie cannot tamper with or corrupt the message. When Bob receives the message, he can trust it is the same one that Alice sent.

- *Availability*: The message is time-sensitive, and Alice wants to make sure Bob receives and reads it. Charlie cannot prevent Bob from accessing the message.

The following sections discuss common attacks and mitigations using this model of confidentiality, integrity, and availability. This structure is useful both in categorizing past attacks to understand vulnerabilities in your system and in thinking ahead to incorporate best security practices into your application. Keep in mind that this list is not exhaustive, nor can all attacks be this neatly bucketed into categories. However, this discussion is a useful example of using the CIA Triad to think about security in the context of an application.

Confidentiality

Everyone has secrets, whether they be embarrassing nicknames or personal relationship history. Like people, all applications have secrets that they want to keep private. The most obvious application secrets are passwords and user data. When a piece of data (such as a Social Security number or birth date) can be used to identify an individual, it is known as Personally Identifiable Information (PII). It is the job of the application to guard both PII and other

---

[5]This framework for understanding cryptography dates back to the 1978 paper outlining the RSA cryptosystem. Alice and Bob, among a slew of other characters, show up in many texts about cybersecurity.

user information from prying eyes. Confidentiality, simply put, is keeping secrets secret—the difficulty of which we all learned at a young age.

One way attackers target confidentiality is through social engineering attacks, which use techniques based in human psychology to manipulate a user into revealing confidential information. The most widespread type of social engineering attack is phishing, the practice of soliciting sensitive information or getting users to download malware by posing online as a credible entity. This section focuses on three types of phishing: drag-net phishing, spear phishing, and clone phishing.

Drag-net phishing attacks cast their sights on a wide swath of users by utilizing mass email schemes. Pretending to be some authority (e.g., the IRS), an attacker sends scores of users the same, seemingly valid email. However, the email includes a malicious link that leads users to a page that looks official or identical to an actual website (e.g., the official IRS website) but is not. Users fall into the trap of entering personal details (e.g., their Social Security number) on this site, enabling the attacker to commit identity fraud. Some phishing emails instead or additionally disguise a destructive virus as an innocent-looking attachment. Users that download and open this file inadvertently cause the virus to run, therefore giving the attacker access to their computers. The most famous drag-net phishing attack is the "Nigerian prince" scam, in which attackers don the guise of a prince that needs help transferring his wealth. The attacker's emails ask recipients to send a seemingly nominal fee to a bank account with the promise of greater wealth if they help. Of course, there is no prince, and the victims can't reclaim the funds they sent to the bank account. While this attack does not always involve the exchange of sensitive information, attackers use the same tactics as many other drag-net phishing attacks.

Spear phishing attacks, on the other hand, target specific victims. An attacker poses as someone the victim knows or trusts and sends them an email that incorporates personalization, such as by addressing the victim by name or job title. A common spear phishing strategy is to email users of a particular service provider—here we'll use the example of Gmail. The attacker, using an account that seems to be an official Gmail account, tells users via email that their accounts were hacked. The email then redirects the users to a fake page that collects their credentials. Since the user thinks they are interacting with Gmail, they are willing to enter their password. This exact attack was the basis of the email leak targeting high-ranking Clinton campaign official John Podesta in early 2016.[6]

---

[6]Lorenzo Franceschi-Bicchierai, "How Hackers Broke Into John Podesta and Colin Powell's Gmail Accounts," Motherboard - VICE, October 20, 2016. Available at https://motherboard.vice.com/en_us/article/mg7xjb/how-hackers-broke-into-john-podesta-and-colin-powells-gmail-accounts.

Clone phishing takes this personalization further and uses the user's previous emails against them. Using a previously received legitimate email as a template, the attacker sends the victim a near-replica of the email, save for a new malicious link or attachment. Since these emails closely resemble actual emails that the user has received, it is more difficult to distinguish between a phishing and a legitimate email.

Email providers are continually incorporating new anti-phishing tactics into their services. Natural language processing, machine learning, and other pattern-matching techniques have been used with great success to find phishing emails and mark them as spam. While the phishing attacks we've discussed so far require some degree of technical literacy, many attackers use nontechnical routes. One such example is voice phishing (also known as vishing), a form of phishing conducted over the phone. In one shockingly successful technique, attackers pose as IRS agents over the phone and threaten the call recipients with arrest unless they make a payment on a supposedly official IRS site. Of course, the money goes to the attacker instead of the IRS.

To protect yourself from phishing attacks, you must stay vigilant. Do not open emails from unknown senders or emails with suspicious subject lines; while spam filters have come a long way, they aren't perfect. Furthermore, it's good practice to stay away from giving out personal information over the phone or email. If necessary to give someone such information, be entirely sure that you are corresponding with the actual authority figure.

In this section, we explored how attackers gain access to view a user's data through the use of social engineering and phishing attacks. Other common confidentiality attacks include packet sniffing, port scanning, and keylogging. When thinking about protecting your confidentiality, consider locking access points to your information (the most straightforward being locking your computer or phone when not in use) including yourself. Next, we examine attacks on a message's integrity.

Integrity

In the grade-school game Telephone, a message is passed around in a whisper from person to person. As you might imagine, it quickly morphs into something completely different as people spread their misinterpretations (which are then also misinterpreted). While not intentional, this game often compromises the integrity of the original message. Integrity here refers to a message's trustworthiness: the message recipient can trust that they have received the message unaltered, from whom they expect. Due to the difficulty of impersonating another secure user, integrity attacks are often technically complex. We'll discuss three kinds of attacks in this section: middle-person, injection, and forgery attacks.

Middle-person, or man-in-the-middle, attacks occur when an attacker intercepts the message between the sender and the receiver. One example is a replay attack, where the attacker Charlie eavesdrops on Alice and Bob to steal a secret, perhaps a password. Then, in a later transaction, Charlie can use this secret to impersonate Alice. In that transaction, Bob thinks the message is from Alice, but the message's integrity is gone since it is actually a different message from Charlie. Replay attacks showcase how attacks can fall into multiple CIA categories; while replay attacks target the message's integrity, Charlie's knowledge of Alice's secret also makes this a confidentiality attack.

Injection attacks involve attackers running destructive code through an application by providing malicious input. For example, consider a web application that uses user input to execute a SQL database query. An attacker might be able to input harmful code, causing the application to run a damaging query. This query could, for example, destroy the entire database. Injection attacks are so prevalent and damaging that the category is listed first in the Open Web Application Security Project (OWASP) Top Ten Most Critical Web Application Security Risks of 2017.[7]

Cross-site scripting (XSS) is another type of injection attack. In XSS, attackers insert code on a website to capture sensitive information, often cookies. Remember that a cookie is a string-based token that websites use to save user-specific settings and authentication. Imagine that your website renders the last result of an HTML-formatted user-submitted comment. The comment might include a script tag with a JavaScript script that steals the cookies of whichever user is visiting the page. When a new user loads the page, this script automatically runs, and the attacker is then able to use these stolen cookies to pose as the victim and commit identity theft.

Forgery attacks include attacks where an attacker impersonates a user. While this also happens in middle-person attacks, those attacks target ongoing exchanges between two parties, while forgery attacks focus on one-way communication channels. For example, cross-site request forgery (CSRF) preys on communication from a user to a website. CSRF exploits the fact that cookies remain active after you exit a website, allowing it to recognize you upon your return. Say that you've visited your favorite shopping site recently; even if you've left the site, it keeps a cookie. An attacker might get you to click a link in a phishing email or run a script through XSS that sends a malicious request (perhaps deleting your account) to the shopping site. Since the cookie has not yet expired, the website may allow this request to execute. A standard way to prevent these attacks is by issuing a hidden session-specific token (called a CSRF token) that only allows requests to run when the user, and not anyone else, is currently on the website.

---

[7]"OWASP Top 10 - 2017: The Ten Most Critical Web Application Security Risks," Open Web Application Security Project. Available at https://www.owasp.org/images/7/72/OWASP_Top_10-2017_%28en%29.pdf.pdf.

As a takeaway, never trust the user, especially their inputs. You should validate their input by requiring that it conforms to a set pattern (e.g., a bank would check that requested deposit and withdrawal amounts are valid, nonnegative decimal numbers). By validating their input, you mitigate the risk of hackers injecting code into your application or otherwise causing harmful outcomes. For many integrity attacks, having a thorough understanding of your particular tech stack (the programming languages and frameworks your application uses) is critical for understanding vulnerabilities where attackers could inject code.

This section discussed perhaps the most technically complex security attacks, integrity attacks. We investigated middle-person, injection, and forgery attacks that seek to inject an attacker's code and cause damaging results. Other integrity attacks include data diddling and salami attacks. To protect ourselves, we learned to rigorously validate our users' inputs and ensure that any assumptions we make are true. Our final section discusses attacks on data's availability.

Availability

Every small, struggling company dreams of the day their application becomes popular and their traffic spikes 100 times. Congratulations—it's happening to you! Unfortunately, your servers aren't set up to handle this much traffic. They crash, bringing down your site. In 2013, this happened on a catastrophic scale: healthcare.gov buckled under the pressure of millions of Americans trying to sign up for the Affordable Care Act. Luckily, a group of engineers saved the site and successfully enrolled millions in healthcare insurance, but not before the underlying infrastructure caused heartache and frustration across the nation. When an application is unable to handle traffic like this, it's known as a denial-of-service. Since resources cannot be reached in such a situation, it is said to compromise the availability of the resources.

Some attackers intentionally cause denial-of-service by sending fake requests to your application until it crashes, leaving legitimate users without access. Such an attack is known as a denial-of-service (DoS) attack. While the only way to combat a legitimate denial-of-service is to expand the amount of traffic your application can handle, you can prevent or counter a denial-of-service attack by blocking destructive requests. There are two main components to a successful anti-DoS plan. The first is to define a set of expected traffic patterns, and the second is to remove traffic that doesn't match these patterns.

A straightforward way of doing this is by tracing the servers originating the attackers' requests and blocking all traffic from those servers. However, this becomes difficult in the most common iteration of a DoS attack, a distributed denial-of-service (DDoS) attack. In a DDoS attack, the malicious requests originate across many servers, making it trickier to create patterns that successfully filter them out. Services such as Cloudflare use advanced pattern creation techniques such as machine learning to protect your application from DoS and DDoS attacks.

Attackers also use ransomware to target availability of data. Ransomware is software that denies users access to their data, sometimes just by locking their computers (annoying but relatively trivial for a security professional to address). More often, however, ransomware extracts and encrypts users' data, after which the attackers demand a ransom in exchange for returning the decrypted data. As we'll learn in the next section, it is nearly impossible to break well-done encryption without its decryption key, so victims must resort to paying the ransom to recover their data.

Availability attacks are especially dangerous for hardware devices that contain code or connect to the Internet. Stuxnet, a computer worm, destroyed hundreds of thousands of centrifuges (lab equipment) by targeting machines running a specific kind of software and forcing the corresponding centrifuges to spin themselves to death. Attacks like Stuxnet fall under control system security, which focuses on industrial systems. While broken centrifuges are replaceable, attacks on control systems can take down large-scale systems like hydraulics or dams, with far-reaching consequences.

This section examined attacks that don't access or manipulate data but instead prevent the intended recipients from viewing and using the data. Other availability attacks include physical electrical power attacks and session hijacking attacks. Validating the requests we receive and filtering out any that seem suspicious is critical for stopping availability attacks.

While the CIA Triad is a useful framework for finding vulnerabilities and brainstorming ways hackers may want to steal your information, it cannot exhaustively account for all attacks. Therefore, it is of the utmost importance to stay in the know about best-practice security protocols by attending conferences such as Black Hat or Defcon, reading papers, and staying up-to-date on the technology capabilities and flaws of the services you use. The US National Institute of Standards and Technology (NIST) and the European Union Agency for Network and Information Security (ENISA) are also excellent resources for information on cybersecurity protection.

Precautions

Luckily, there are precautions that can help you secure your application. While these widely accepted security practices go a long way, they aren't comprehensive and this section serves more as an overview. You should have a security team dedicated to strengthening your security practices and dealing with attacks.

Encryption

One of our best defenses is encryption, the practice of transforming a message from a "plaintext" into a "ciphertext" so that a third party cannot figure out the plaintext from the ciphertext. At its core, encryption is just changing the presentation of a piece of data, meaning all data types (including audio and images) can be encrypted. A straightforward, not very secure, method of encrypting a plaintext is to reverse the characters. For example, we could take the plaintext "the quick brown fox jumps over the lazy dog" and come up with the ciphertext "god yzal eht revo spmuj xof nworb kciuq eht." It looks like a garbled mess, but a keen eye might notice two occurrences of "eht," relate that to the common word "the," and figure out both the encryption algorithm and the secret message.

This example shows how a simple encryption schema is insecure. Since mathematical and linguistic models can crack simpler encryption protocols, you should use well-known and rigorously tested encryption protocols and systems. While more sophisticated encryption methods could theoretically be cracked, the magnitude of resources and time needed to break them render them practically unbreakable.

The basis of many encryption systems and protocols is public-key cryptography, which uses two auxiliary pieces of information: a public key and a private key. A private key must be kept secret by its owner, while anyone can access the public key. If Alice wants to send Bob a plaintext message encrypted using public-key cryptography (also known as asymmetric cryptography), she inputs her message and Bob's public key into an encryption algorithm, which then generates a ciphertext. Upon receiving the ciphertext, Bob inputs the ciphertext and his private key into a decryption algorithm to decrypt the message. Creating public and private keys in a way that is both secure and efficient is difficult. Key generation cryptosystems (RSA being the most well-known) often employ math to create large numbers that act as keys.

TLS (Transport Layer Security), used to secure communications over the Internet, is one such encryption protocol based on public-key cryptography. It has two parts: the first authenticates the two parties (a client and a server) and creates a secure channel to exchange messages, while the second provides security for the data moving from client to server or server to client. TLS is used ubiquitously across the Internet to secure websites, APIs, and more.

Encryption is crucial for keeping data secure. All data, including both customer and company data, should be encrypted and only be decrypted when necessary. The life cycle of a piece of data has three parts: data in use (data pulled out of permanent storage), data at rest (data stored in persistent memory), and data in transit (data in between two systems). We use encryption to secure data that are at rest and in transit—for example, TLS protects data in transit.

In the communication and messaging field, some products have taken care to encrypt data throughout its life cycle, a feature known as end-to-end encryption. Let's think about this in the context of Alice and Bob. If a message Alice sends Bob is end-to-end encrypted, the original message is only ever visible to Alice and Bob. Even if Alice uses a third-party client to send her message, the application discards the plaintext as soon as it is encrypted.

We mainly discuss end-to-end encryption in the context of peer-to-peer communications, like email and messaging services. End-to-end encryption has relatively low usage in email primarily for two reasons. The first issue is our reliance on email providers to filter spam, meaning the providers must be able to read email contents. The second is the fragmentation of email providers and lack of industry standards for end-to-end encryption. Sending encrypted email from, say, a Gmail account to a Yahoo account, would require both providers to align on one encryption schema or users themselves to align on an encryption schema and exchange public keys, which is burdensome. However, end-to-end encryption is slowly becoming ubiquitous throughout messaging. To remain truly secure, you should use the Signal messaging application, which both supports end-to-end encryption and is open source (meaning anyone, including security professionals, can and have verified its security by reading its source code). The platform is so secure that other (closed-source) messaging applications such as Facebook Messenger and WhatsApp use the Signal Protocol to implement their end-to-end encryption.

Password Protection

On an individual level, the most important precaution is to use strong passwords. There are many best practices for creating secure passwords: using special characters and numbers, increasing the password length, using a passphrase, never reusing a password, and more. While these best practices are, admittedly, easy to let slide, they are crucial to follow: a 2016 Verizon data breach report found that 63% of data breaches are a result of weak or stolen passwords.[8]

The 2017 report "The Password Exposé" by password manager LastPass showed that the average LastPass business user has 191 passwords, far too many for any one person to remember.[9] Password managers like LastPass and 1Password make organizing and using passwords much easier by storing all

---

[8]"2016 Data Breach Investigations Report," Verizon, April 2016. Available at http://www.verizonenterprise.com/resources/reports/rp_dbir-2016-executive-summary_xg_en.pdf.

[9]"The Password Exposé," LastPass, November 1, 2017. Available at https://lp-cdn.lastpass.com/lporcamedia/document-library/lastpass/pdf/en/LastPass-Enterprise-The-Password-Expose-Ebook-v2.pdf.

of your passwords in a "vault" protected by one master password. Password managers also have an Auto-Fill feature to fill in your credentials on websites and a Generate Password feature to automatically create strong passwords for you. This way, the only password you need to remember is a single master password. The idea is that it's much easier to remember one complicated password than many passwords, however memorable they may be. Therefore, all of your passwords can be much more difficult to guess and thus stronger.

While password managers are not immune to hackers, their data sit in a vault secured by several layers of encryption. Even if hackers steal a password manager's vault, they cannot access your data because it sits behind this encryption, locked with your master password as a key. To decrypt the vault, they would need your master password (which isn't stored anywhere, thanks to a process called hashing). As long as you've chosen your master password wisely, the hackers won't be able to guess it. In short—while switching to a password manager might seem scary, the majority of security researchers agree that it is the safest way to create, store, and use passwords.

Once a user creates their password, it's crucial that the password is never exposed. As mentioned above, we use the process of hashing to protect passwords. Like encryption, hashing converts a piece of data into a different form, here called a hash. Unlike encrypted ciphertexts, hashes are not reversible. They work well to verify information because the same hashing function run on identical copies of a plaintext (say, when a user first chooses their password and when they enter their password on the site) produces the same result. However, we do not store the plaintext and thus cannot retrieve it from our hashed copy.

Means of Authentication

Factors of authentication are artifacts that a user must have to prove that they are who they claim to be. There are three types of authentication factors: knowledge factors ("something you know"), possession factors ("something you have"), and inherence factors ("something you are").

Knowledge factors are the most common means of authentication. For example, a password is a piece of information that a user must *know*. Another example of a knowledge factor is a security question ("What is your mother's maiden name? What's the make and model of your first car?"). As anyone who has seen the film *Inception* knows, knowledge factors are relatively easy to crack for three reasons: (a) they're often simple, (b) their information can be fixed, meaning that it never changes, and (c) they follow a familiar format. Since humans must remember their passwords, they are often inclined to base them on personal information like a birthday or a pet's name that tends to be easily guessable. Security questions surface information that some sleuthing could uncover, and even if an attacker can't find the make and model of your first car, there are only so many possible answers that they would need to try. While knowledge factors are currently the most prevalent form of authentication, they don't provide the security that is increasingly necessary.

Therefore, many systems are moving toward using two-factor authentication (2FA), which requires another layer of authentication on top of knowledge-based authentication. The second layer is usually a possession factor, which authenticates a user based on an object that they have. Think of the keys you use to unlock your home and car. For many web applications, the possession factor is the user's mobile phone. The application provides the mobile phone with a one-time password (OTP), either by SMS or through a mobile application such as Authy or Duo. SMS can be hacked more easily since it is tied to your phone number. Attackers use phishing attacks where they call a cell service provider under the guise of losing their phone. The provider allows them to switch their number to a new phone, enabling them to steal the OTP. The most secure possession factor is considered to be a separate hardware token built explicitly for this purpose such as a Yubikey. The Yubikey is a small device that plugs directly into a user's computer and generates OTPs without using the Internet.

Sometimes, the extra layer uses an inherence factor instead. An inherence factor is a biometric measure, which utilizes a physical attribute of the user. Touch ID on iPhones, which allows users to log in with their fingerprint, is an inherence factor, as are retinal scans and facial recognition. While inherence factors can be easy to use (scanning your fingerprint may be more convenient than typing in an OTP), they are not as popular as other authentication schemas because a compromised biometric is impossible to replace. If, for example, hackers manage to steal your fingerprint data, there is no reasonable way for you to change your fingerprints, whereas a password is simple to change. This risk is allayed in most uses by storing biometric data locally (e.g., directly on your phone) under encryption. Since the data aren't traveling between locations, the attack surface (the number of different places an attacker could target the application) is smaller, meaning there are fewer ways for an attacker to hack the application. Therefore, inherence factors generally only make sense for applications tied to a device that doesn't need to store biometric information in the cloud.

Multifactor authentication (MFA) can protect you against attackers who have stolen your credentials, say from a replay attack. If a hacker tries logging in with your password, they won't be able to log in since they don't have the second layer of authentication. Keep in mind that you should still rotate your credentials if you suspect someone has stolen them. For personal use, the most secure decision is to add MFA on all applications that support it, from your bank account to your social media accounts. However, requiring MFA for your users may not be the right decision. While MFA provides additional security, not all consumers prioritize security and may consider MFA burdensome. In any case, it's always a good idea to support MFA, even if you don't require it.

Network Security

Network security refers to the securing of networks that access and transport information, both from your user's side and in a corporate setting. This section outlines tools you can use as an individual user and for everyone on your Internet network, whether that be a home network you share with your family or a corporate one you share with your company.

From a personal standpoint, the simplest precaution you can use to protect yourself while browsing the Internet is using HTTPS. HTTPS, which stands for Hypertext Transfer Protocol Secure, uses TLS (from the encryption section above) to serve traffic safely. While browsing the Web, make sure that you use the HTTPS protocol at the beginning of your URL instead of HTTP, which does not include the encryption that HTTPS does. In some browsers, like Firefox and some versions of Chrome, you can also verify that you are surfing with HTTPS by looking for a green lock in the URL bar. Browsers may also naturally block connections to websites using outdated or unsafe TLS settings.

Beyond using HTTPS, make sure to exercise caution on the Web. Download software only from trusted sources, and be mindful of streaming sites with ads that might contain malware. A simple way to combat such ads is to install an ad blocker such as uBlock. Since attackers often create hyperlinks that claim to link to one domain (e.g., "google.com") but send you to a different, dangerous site, it is good practice to type URLs into the browser instead of following links to protect against the phishing attacks we outlined earlier.

To further secure your web traffic, you can use a Virtual Private Network (VPN), usually by using a third-party VPN service. A VPN sends network requests (most frequently website navigation) through an encrypted tunnel to a remote server, which then sends the request on to the public network. When browsing via a public network (like at a coffee shop), this is useful because your traffic is encrypted and appears to be coming from a different IP address owned by the VPN service. Your data are still secure even if the public network is insecure because the VPN tunnel encrypts them.

Many workplaces, especially those with remote workers, utilize a VPN so that employees using different local networks (say, their home Wi-Fi networks) appear to be on the same private network. Then, computers on the VPN can identify which other computers are on the same VPN and are therefore trustworthy. In essence, a VPN can serve as a single point of authentication, much like badging into an office space.

A final precaution you can take to secure any network is to use a firewall or a proxy. A firewall blocks traffic from user-specified sites to your IP addresses and vice versa to protect you from malicious sites and attackers. A proxy serves as an intermediary between the user and the external network and can perform various functions, such as blocking malicious sites and masking your IP address to any prying parties or potential attackers.

Implementing Security Practices in Your Application

Given all of these attacks, tools, and precautions, the next step is to take this knowledge and use it to build security into your team's development process. We'll outline three steps to guide you in thinking about security as it pertains to your application: (a) securing your application, (b) staying secure, and (c) minimizing damage.

Securing Your Application

It might be overwhelming to think about securing your application. Luckily, several open-source libraries and packages implement protocols and best practices for you. Using libraries is essential because security is difficult to implement—and poorly implemented security leaves your application wide open to attacks. Popular open-source libraries tend to have robust implementations since anyone can review the code and point out loopholes. Look for libraries that are well-maintained and whose developers respond to issues promptly.

Many of the attacks and mitigations we've discussed come back to the same lesson: never trust the user. In particular, several integrity attacks, such as SQL injections, rely on the ability of bad actors to provide input that causes the application to perform harmfully. Validating user input as discussed earlier is one piece of protecting yourself against these attacks.

A second piece is access control, which involves establishing permissions so that each user has access to only what they need. Consider a bank that gives each user a homepage at mybank.com/username/home. By guessing usernames, attackers could find other users' homepages and access sensitive information in a clear breach of confidentiality. Instead, the bank can check that the username of the logged-in user matches the username associated with the page they are trying to access before loading the content. Proper access control means that an application ensures that a user has the right permissions before they view the content (e.g., their bank homepage) or perform an action (e.g., withdraw money from an account).

Staying Secure

Now that you've built a secure first version of your application, you might wish to focus all of your efforts on marketing your product. Unfortunately, as the old saying goes, "ain't no rest for the wicked"—bad actors don't stop trying to attack your resources. Even when you are not actively developing new features, you must take precautions to ensure your application stays secure—two essential measures are testing and installing updates.

Security testing builds on top of the testing done while developing the application. Routine testing won't necessarily catch all bugs. Finding security bugs is of exceptional importance since they can lead to dire consequences, while logic bugs or rendering bugs lead to poor user experience at worst. To catch these high-risk bugs, developers use penetration testing, a model that simulates a security attack on the application. Penetration testing often employs the use of fuzzing, the practice of randomly generating inputs to an application and confirming that the outputs are both valid and expected. Fuzzing is useful for simulating edge cases that other tests may not have otherwise uncovered. Many companies also employ ethical hackers to find flaws in their system so that they can fix them before bad actors find and exploit them.

Your application's bugs aren't the only ones that are cause for concern. It's inevitable that your work will rely on platforms, frameworks, and libraries built by others. Sometimes, one of these third parties discovers a security flaw and releases an update to fix it, also known as a patch. Promptly installing updates is crucial; until you install the patch, your application remains vulnerable to the security flaw it is supposed to fix. In the case of the Equifax breach, hackers exploited an old version of a third-party library that Equifax had neglected to update, months after a patch was released.[10] These vulnerabilities exist in even the most established libraries—Heartbleed, a bug that opened the way for one of the most destructive attacks ever, was part of the OpenSSL library, widely used to execute the TLS protocol. As we discussed, TLS is one piece of how computers talk to each other. Hackers exploiting the Heartbleed protocol could manipulate requests to trick a computer into sending more information than they had access to; sometimes, this data included personal information such as usernames and passwords. Here, the crucial discovery for hackers was realizing that the receiving computer didn't validate the amount of information that they were requesting. Once again, this example illustrates the need to validate user inputs.

Minimizing Damage

Even if you follow all of the best practices described above, a hacker could still break into your systems. New methods of attack emerge all of the time, so an application is never 100% secure. After an attack, you must work to secure your resources and to recover as quickly as possible. The first step is to check your system logs. A log is a report of some action, transaction, or event. Computer logs are useful because they can continuously monitor a system, and you can choose what to include in them. For example, you might

---

[10]Lily Hay Newman, "Equifax Officially Has No Excuse," WIRED, September 14, 2017. Available at https://www.wired.com/story/equifax-breach-no-excuse/.

log when a user signs in or when your application makes an API call. Most of these logs will and should be innocuous, as most of the activity on your site should be valid. However, when there is suspicious activity, logs can make it easier to spot malicious behavior or fraud. Take the case of a credit card company that notices a user's credentials popping up around the world. This could be a world traveler—or a stolen identity. Either way, logs provide insight into the situation and how best to move forward. You may need to fix a bug or build in more precautions, like adding offending sites to a blocklist.

After being struck by an availability attack and losing access to your data, you need to restore services as soon as possible. For this reason, it is a best practice to maintain encrypted backups of your data in different physical locations. Even if you lose data accidentally (perhaps you drop your phone in water), backups ensure that you can securely restore your data in some capacity, instead of losing everything. Storing backups in different locations protects data from physical attacks.

Backing up data is crucial, particularly when you provide sensitive services. Let's examine the case of the WannaCry ransomware, which hit over 60 organizations in the National Health Service (NHS) of the UK.[11] While the NHS had lax security practices that allowed WannaCry to breach their systems, their backup system allowed the majority of organizations to recover from the attack with no data loss and without paying the demanded ransom. The ability to recover quickly was vital for a health system serving critically ill patients. As a result of robust enterprise backup systems, WannaCry was a much less financially successful attack than other ransomware because affected businesses only lost a copy of their data.

Conclusion

In this chapter, we've touched on some of the most widely known attacks and mitigations in the cybersecurity realm. We established the CIA Triad as a model for thinking about security and examined ways that attackers might poke through your safeguards. Furthermore, we discussed possible protections such as encryption, multifactor authentication, and firewalls. Finally, we outlined security as a process, starting with building in security and ending with minimizing damages from attacks. Unfortunately, your application is a target for bad actors, and it is your responsibility to protect yourself and your company against them. Good luck building your defenses!

---

[11]Nicole Reineke, "Will Ransomware Start Targeting Enterprise Backups," Unitrends. Available at https://www.unitrends.com/blog/will-ransomware-start-targeting-enterprise-backups.

Mobile Basics

Special thanks to Ashi Agrawal and Cheenar Banerjee for contributing this chapter.

Until now, we've assumed that MyAppoly exists solely as a web application, but the reality is that many products—think Snapchat or Lyft—live partly or entirely in mobile form. These companies optimize their application experience for your phone, a small device that moves around with you and comes equipped with a different set of tools than a traditional computer. Many of the structures we've discussed thus far are still relevant, but a successful mobile application does more than recreate the desktop experience: it is built specifically for mobile devices. So, let's dive into the intricacies of mobile development!·

A Short Introduction

Ever since the invention of the cell phone in the 1970s, the capabilities of the handheld computer have rapidly grown. Blackberry devices rose to prominence in the early 2000s because of their abilities beyond SMS texting and calling, including a focus on email. Then, with the advent of smartphones, the number of things a user could do exploded beyond just making phone calls. Now, we track our fitness, set alarms, and call ride-sharing services from these incredibly convenient devices. This chapter focuses on smartphones, but mobile devices also include handheld computers.

© Vinay Trivedi 2019
V. Trivedi, *How to Speak Tech*, https://doi.org/10.1007/978-1-4842-4324-4_13

Mobile applications constitute a booming market. The number of global mobile phone users passed the five billion mark in 2018.[1] Among websites, mobile devices drive more traffic than desktop devices,[2] and mobile applications are expected to drive nearly 200 billion dollars in revenue yearly by 2020.[3] That's exciting, but what *is* a mobile application? It's a program intentionally designed and written to run on a mobile device. As we'll see, mobile applications can open doors but come with new considerations and limitations.

Why Develop for Mobile

Before diving into *how* to develop a mobile application, you need to understand *why* and *when* to develop one. In some cases, this is obvious—for example, ride-sharing and mapping applications have clear mobile use cases that are crucial for success. It's not quite so cut and dry for other products, but mobile often makes sense in a roadmap because it can diversify and augment functionality and consumer base. In the case of MyAppoly, your team decides that a separate mobile presence is necessary to attract new consumers.

While MyAppoly has built a desktop web application first, some applications consider mobile-specific features and constraints the default design. This process is called mobile-first development and leads to a better mobile user experience and a cleaner, more minimal desktop design. A mobile design fits fewer elements on the screen, which translates easily to the desktop, but moving from desktop to mobile means that designers have to think about what to remove.

Mobile Apps vs. Desktop Apps

Mobile apps are quite similar to the web applications we've been discussing. Like traditional web apps, mobile apps have a front end and a back end. Both also utilize APIs and other external resources. In this chapter, we'll refer to traditional web apps as desktop web apps. There are also some desktop apps that you download, like a word processor or movie editing software, but this text primarily compares mobile applications with desktop web applications.

---

[1]McDonald, Nathan, "Digital in 2018: World's Internet Users Pass the 4 Billion Mark," *we are social,* January 30, 2018.
[2]Enge, Eric, "Mobile vs Desktop Usage in 2018: Mobile takes the lead," Stone Temple, April 27, 2018.
[3]Takahashi, Dean, "Mobile app market to grow 270% to $189 billion by 2020, with games accounting for 55%," VentureBeat, November 2, 2016.

There are significant differences between developing for mobile devices and desktops. Let's outline these differences to understand why we need to create a separate mobile presence. There are four main differences between the two platforms: size, method of interaction, features, and portability. Let's dive into each.

- *Size*: Mobile device screens are much smaller than desktop screens, which restricts the amount of content per page. The user interface of mobile apps must be adjusted accordingly. For example, many mobile apps feature collapsible menus to save screen real estate, while desktop web apps display large menu bars to lay out all of the options.

- *Method of interaction*: We also interact differently with our mobile devices than with our desktops. While we use a mouse or trackpad to navigate around a desktop screen and a separate physical keyboard to type, we use touch with modern smartphones.

- *Features*: Mobile devices come equipped with a different set of native, or built-in, features than traditional laptops do. Often, this includes a camera, location services, Bluetooth, and a push notification system. As a result, there's a whole different set of functionalities that we can create for mobile apps, for example, Instagram launched on mobile devices because most people primarily use their phones to take photos.

- *Portability*: Portability means that people take and use their smartphones everywhere. As a result: (a) the presentation of content needs to be simple, (b) there are additional real-time use cases, and (c) cellular data is different than broadband Internet. Let's investigate these in further detail. People often browse their smartphones for short intervals of time, so user interactions need to be snappy. Layers of menus offer more choices in desktop web apps but aren't practical for a person on a time budget. Since people check their phones on the go, mobile applications can focus on real-time actions, like ride sharing. Finally, applications often access the Internet via cellular data, which can be slower and less reliable than Wi-Fi or Ethernet. As a result, mobile applications might be optimized for speed. With so many mobile apps competing for users' time, users may switch apps instead of waiting for a slow app.

As a result of these differences, we must think critically about not only the restrictions that mobile development brings but also the expanded possibilities and opportunities when creating mobile content.

Types of Mobile Apps

There are three main types of mobile apps: web apps, native apps, and hybrid apps. All three are common, and the best kind of app to develop depends on the product itself. This section discusses the technical underpinnings and user presentation of each type.

Web Apps

Mobile web applications, web applications that the user accesses by opening a URL in a web browser like Safari or Chrome, are the closest to desktop web applications. They often go by a more familiar name: mobile websites. For many products, these applications simply use a different front-end wrapper around the same back-end powering the desktop web application. Other products use one responsive website for both desktop and mobile devices. Responsive websites change based on the screen size of the device, which could also be a tablet or a rotated mobile device. Elements on-screen automatically scale, hide, or move to provide a good user experience across different devices.

While mobile websites are often relatively easy to develop on top of a desktop website, they run relatively slowly because they run within a browser rather than directly on the device itself. Furthermore, they have limited access to the native features that often make mobile apps so powerful. As a result, many companies have started to make standalone mobile applications.

Native Apps

A native app is written to run directly on a phone's operating system. The two most popular operating systems are Android and iOS, which are developed by Google and Apple respectively. While some phones run Windows or Linux-based operating systems, the vast majority run either Android or iOS. Native applications are usually downloaded via an app store and appear as icons on the phone screen. Because native applications run directly on the phone, they do not use a browser intermediary and therefore run more quickly than mobile web apps do.

A prominent reason that developers write native applications is the ease of utilizing additional features, including the camera and GPS. These features, unlike an external back end or API, can be accessed offline and allow the application to have offline capabilities. Native apps can also use local device storage to store an offline database that persists between different uses. These applications track and save user interactions and don't need Internet service to serve content, such as locally stored photos or notes—think of a gallery app or a music player.

Additionally, this local database can be useful in serving personalized content. While desktop web applications can store information for personalization, they generally use user accounts, like linking all of your past purchases together in a rewards membership. Since people use multiple browsers or computers but use just one phone, a user account for a mobile application is usually just the set of interactions that take place on that particular phone. As a result, many mobile applications can remove mandatory sign-in, which is often a barrier to entry for new users.

A final note about native apps is that they have more built-in security precautions than web applications. Mobile operating systems provide additional security measures to standalone apps that they cannot give to mobile websites. Apps also have to pass a review process to qualify to be available for download from the app store in the first place, while there's no such verification system for mobile websites.[4] However, native mobile applications can provide a more valuable target to attackers since they have access to more features than mobile websites do. Both mobile apps and mobile web apps are susceptible to security breaches, so it pays to be mindful regardless of the route you choose.

Hybrid Apps

The final category of mobile applications is hybrid apps, which combine aspects of native and web apps. Like mobile websites, hybrid apps run using a web browser, though here the browser is built into the app using a web view. As a result, hybrid apps are not as fast as native applications. However, they have some native benefits, like access to device-specific features such as the camera and GPS. Furthermore, hybrid apps are downloaded from an app store and appear as standalone icons on phone screens. Some hybrid applications are written using desktop web technologies, meaning they can be easier to write if a desktop web app already exists.

While hybrid apps can leverage native technologies and can be easier to write, they come with a different set of drawbacks. Using a separate web view is slow and doesn't allow access to all device features. Additionally, since these web views use web technologies, their designs often don't comply with device design rules. Even when the web views abide by these rules, they can be difficult to use. Companies often create hybrid applications earlier in their

---

[4]See https://developer.apple.com/app-store/review/.

development cycle to spin up something lightweight and then switch to a native application in the long term. Hybrid apps are also popular options for teams with web development experience but no mobile platform-specific expertise or for simple applications that can sacrifice speed.

Now that we've discussed the high-level differences between web, native, and hybrid apps, we'll discuss development for standalone applications (native and hybrid apps), since they are written using different technologies from desktop apps.

Standalone Application Development

Standalone applications refer to applications that are accessed via their own icon instead of by web browser. These applications can be written specifically for the operating system they run on or more generally for all smartphones. This section outlines each approach and the relative benefits and drawbacks.

Android vs. iOS

As discussed earlier, native mobile apps run directly on a phone's operating system. Developing for both Android and iOS is a hefty task, so many companies focus on building out one platform first. Choosing which platform to focus on can be challenging, so here we'll outline the key differences. Android and iOS differ primarily in three ways—user base, development process, and app deployment.

For most products, the primary consideration in choosing a platform is the consumer base. While iOS has a stronghold on Western consumers, Android has more users globally. Due to the high price point of the product and its branding, iOS has attracted consumers that spend more money on applications and in-app purchases.

Of course, they are both different to develop for in a technical capacity as well. Android apps use Java or Kotlin, developed in the Android Studio IDE, while iOS applications use a programming language created by Apple called Swift, which is developed in the XCode IDE. Java enjoys widespread popularity, while Kotlin and Swift are relatively new entrants with more limited usage. It may be more likely that one of your teammates already knows Java. Furthermore, XCode only runs natively on Mac devices, meaning the development process is gated by the type of your machine, while Android Studio runs across all types of computers. Because of Apple's policies, development for iOS application development is more restricted. On the other hand, Android is open source, and developers can inspect the source code to understand the platform on a deeper level.

Android applications take more time to develop because of the variety of devices and operating system versions running. There are relatively few devices that run iOS, and they're all manufactured by Apple. As a result, there is minimal variability between them. Android devices, on the other hand, are produced by over 1000 different companies.[5] This variation means that developers have to consider a vast spectrum of screen sizes and device specifications when creating Android apps. Since Google does not manufacture all Android devices, it cannot force users and device manufacturers to push the latest operating system on consumers. Updating operating systems can be a pain for the device manufacturers, who have to ensure it plays well with their phone configurations. Therefore, they often do not push latest releases to consumers, resulting in mobile device fragmentation: Android users split between operating system versions. Android developers thus have to make sure that their app functions well across many versions of the operating system. This outdated software can also lead to increased security risk, as we discussed in Chapter 12. Many hackers take advantage of Android device fragmentation to target malicious code at Android devices.

Finally, Android and iOS differ when it comes to deploying the app on their respective app stores. While the two don't release statistics on the number of app applications they receive, Android anecdotally seems to accept a higher percentage of apps into its app store than iOS. Apps are therefore more likely to be rejected from the Apple App Store than the Android Play Store.

Both Android and iOS come with advantages and disadvantages. When deciding which platform to focus on, products often weigh consumer base and ease of development most highly. If you still can't pick one or want to release to both platforms at once, cross-platform development may be the appropriate route.

Cross-Platform Development

Although truly native apps offer the best performance, it takes quite a bit of development overhead to write and to maintain them for both Android and iOS. Not only does each platform require a different codebase, but each change to the app requires separate updates and testing. As a result, cross-platform solutions are rising in popularity. Cross-platform mobile development refers to a process where the developer writes a single app to run on multiple mobile operating systems. Several existing frameworks support cross-platform mobile development, many of which allow developers

---

[5]"Android Fragmentation (2015)," OpenSignal, https://opensignal.com/reports/2015/08/android-fragmentation/.

to code entirely in JavaScript and other web technologies. These frameworks can be used to create either native or hybrid web applications.

Native mobile applications are written using frameworks that offer developers the use of generic components that are translated on each device into native components. Popular tools include React Native, Flutter, and Xamarin. There are differences in how the three translate platform-agnostic code into native code, but the end result is mobile apps that run almost as fast as completely native apps. Depending on the desired functionality, these applications may still require platform-specific code. However, the majority of the codebase remains the same.

Hybrid cross-platform mobile frameworks include Ionic, Apache Cordova, and Onsen UI. Remember that hybrid applications combine the ease of writing desktop web applications and the features of native mobile applications. Like frameworks for native mobile applications, these frameworks translate platform-agnostic code into hybrid code. Furthermore, these frameworks provide the developer with access to native features, as described earlier.

Native vs. Cross-Platform Development

For many products, cross-platform development makes sense due to the plethora of advantages across three axes: development efficiency, developer accessibility, and user exposure.

The most marked difference between native and cross-platform development is the drastic efficiency improvement across development cycles. Since developers only need to write and to maintain one codebase, releasing new updates is much quicker. The use of web technologies can lead to faster development as well. Native code needs to be compiled before it runs, while web code can instantly refresh (called "hot reload") with new changes. Though compile times range from a few seconds to tens of minutes, these slowdowns can add up over time.

The use of web technologies also makes cross-platform development accessible to more developers. Since fundamental web technologies such as JavaScript are known relatively widely, new developers can ramp up on the project more quickly. Furthermore, cross-platform solutions eliminate the need for the company to hire iOS and Android experts separately for each codebase.

Finally, cross-platform development is an excellent tool for driving user adoption. By employing a cross-platform development process, you can launch to both Android and iOS users at the same time, rather than having to focus on one or the other at the start.

Cross-platform development has its drawbacks, and many prominent companies such as Facebook and Airbnb have turned away from it. It can speed up the development cycle, but the speed of the mobile applications themselves suffers. The device has to translate the cross-platform code into native code, which takes additional time. For applications with time-sensitive use cases that need to run quickly, this can ruin user experience. Furthermore, apps developed via cross-platform development cannot take complete advantage of the native code structure. While these frameworks cover most tools for each platform, they can't capture all of the per-platform intricacies. As a result, more complex products still require somewhat significant amounts of platform-specific code. In the end, this can feel like supporting three platforms instead of two, taking away from the fast iteration speed that cross-platform development boasts. Finally, since cross-platform solutions are still relatively young, they are more likely to provide a buggy experience to developers as they develop into a mature framework. Cross-platform development can serve a fledgling product well, but it is by no means a panacea to the troubles of mobile development.

Conclusion

In the end, choosing when to focus on desktop and when to focus on mobile depends on your product roadmap. Many products want to launch quickly, so they use a cross-platform solution to create a hybrid application, while some products focus on optimizing user experience and create a fully native application. For other products, a mobile website suffices. Deciding which route to choose depends on your bandwidth and your priorities. While creating a meaningful mobile experience for your users requires extra investment, it can drive user growth and satisfaction.

The Internet of Things

Special thanks to Cheenar Banerjee for contributing this chapter.

As you're working on MyAppoly, you realize that your laptop isn't the only thing around you that is connected to the Internet. Your smartphone, smart thermostat, and smart light bulb are just a few of the devices around you that are online. In fact, it's projected that by the year 2020, 50 billion devices will be connected to the Internet.[1,2] There's a technical term for this phenomenon. It's called the *Internet of Things*, or IoT.

Things in the Internet of Things

So what exactly is a Thing in the Internet of Things? It's any device that connects to the Internet and has an embedded sensor. Is a smart light bulb an IoT device? Yes. How about a smart speaker? Yes. Smart toothbrush? Yes. This book? Probably not.

---

[1]"A Guide to the Internet of Things," Intel, https://www.intel.com/content/www/us/en/internet-of-things/infographics/guide-to-iot.html.
[2]Karen Tillman, "How Many Internet Connections are in the World? Right. Now.," Cisco Blogs, https://blogs.cisco.com/news/cisco-connections-counter.

© Vinay Trivedi 2019
V. Trivedi, *How to Speak Tech*, https://doi.org/10.1007/978-1-4842-4324-4_14

Technically speaking, anything that connects to the Internet, including a traditional computer, counts as an IoT device. In a practical sense, though, IoT refers to devices that you wouldn't traditionally expect to connect to the Internet—such as thermostats and refrigerators.

A useful IoT device does the following three things:

1. Collects important data via its embedded sensor(s)

2. Communicates the data over the Internet and receives results of analysis on the data

3. Motivates meaningful action using the results of the data analysis

So, an IoT device has three responsibilities: collect, communicate, and motivate. Let's dive deeper with an example. Suppose the MyAppoly office has just invested in the newest smart refrigerator on the market, created by a manufacturer we'll call Roost. The fridge looks like a regular fridge, except for a touchscreen console on the door. This console allows MyAppoly employees to connect the fridge to the office Wi-Fi, customize fridge features, and view alerts and other information. One exciting new feature of the fridge is its ability to detect when a particular item of food needs to be reordered by using embedded sensors that weigh the food in the fridge. For example, one set of sensors is responsible for weighing the milk shelf.

The milk sensors collect data about the weight of the milk shelf, which the fridge then communicates to another computer via the Internet. The fridge receives an alert when the weight of the milk shelf is low, meaning that the milk carton is almost empty, and motivates action by either encouraging the user to replenish the milk (by console display, email, or text alert) or by ordering a new milk delivery online.

The Internet of Things

Now that we know what an IoT device is, let's talk about a collection of IoT devices working together. The Internet of Things is a network of IoT devices that share data and communicate with each other, without any human interaction required. With so many devices talking to each other, it is important for us to be able to manage their data and communications. We can do that using a piece of software called an *IoT Platform*. Each IoT device in a network connects to the IoT platform, which bridges, stores, and analyzes data from all devices on the network and communicates with each network device individually. The IoT platform exists in the cloud, and all data analytics happen in the cloud. While humans may interact with the IoT devices themselves, in an IoT network, no human interaction is required for the data collection sharing between the devices.

In our fridge example, one IoT network could be the network of all Roost smart fridges in the United States. The platform would be the software that talks to each smart fridge and collects and analyzes data from all the smart fridges on the network. Another IoT network could be a network of smart devices within the MyAppoly office (such as the smart fridge, smart doorbell, and smart speaker). A single platform would still communicate with and analyze data from all of the devices, even though they're all devices with different functions. Amazon Web Services IoT, Microsoft Azure IoT, Google Cloud IoT, and IBM Watson are all examples of existing IoT platforms.

The real power of the Internet of Things stems from the massive amounts of shared data on the network. Let's think back to the MyAppoly fridge, which is part of the national Roost fridge network. The IoT platform for this network collects hundreds of thousands of milk weight measurements every day, from fridges all across the country. Over time, using data analytics and machine learning discussed in Chapter 15, the software can find patterns and insights in the data. One such insight is the optimal milk weight at which a fridge user should order new milk—by looking at large amounts of milk weight data over time, the software can see how long, on average, it takes for new milk to arrive or for users to finish off a half-empty gallon of milk. Insights like this work together to yield a magic milk weight number that is continuously improving as more data come in.

Data from different sources can be combined and analyzed to create equally useful insights. In the MyAppoly office IoT network, an employee could ask the smart speaker about the milk supply, and the speaker could respond with the exact weight of the milk shelf. The speaker gets the information by communicating with the fridge through their shared IoT platform. The fridge and doorbell could coordinate milk delivery orders; over time, based on doorbell data, the fridge could learn only to place orders that will arrive when someone is available to answer the door.

Data and insights from the Roost and MyAppoly office IoT networks could be useful to other businesses, organizations, and IoT networks as well. For example, Roost might learn that their customers tend to drink more milk during the springtime. Roost could sell this data to dairy farms so that they could then manage their production to meet demand more accurately. Alternatively, the office doorbell could notify an employee's home oven to heat up dinner before she gets home in the evening.

Interoperability among IoT systems refers to the ability of IoT systems and devices from different manufacturers to communicate with each other. For example, if the MyAppoly office got a new GE smart light bulb, interoperability would mean that the GE light bulb and the Roost devices could function as part of the same IoT network. Interoperability is encouraged by what are known as open standards. Standards are sets of specifications and rules that manufacturers follow when creating products. Open standards are publicly available, while closed standards are often proprietary. When several companies follow a single set of open standards, the resulting products are compatible with each other. In this case, the GE light bulb and Roost fridge on the same network would be following the same set of open standards. Open standards encourage innovation and make it easy to connect devices from different manufacturers. Closed standards, on the other hand, can lead to more secure and predictable networks, since a single company retains complete control over the standards.

As we can see, our total fridge experience improves when we make intelligent use of its data and data from other fridges. This fridge example summarizes one of the core ideas of IoT: the more information we have about something, the more knowledge we have and the better actions we can take. We'll now talk about the different factors involved in the recent and rapid growth of the Internet of Things.

Factors Driving IoT Growth

The central idea behind IoT has been around for quite a while. One could argue that the first mention of this idea was as early as 1843 when scientists hoped that the collection of machine-measured data about the weather could lead to better predictions in the future.[3] In 1982, a Coke machine was given Internet connectivity at Carnegie Mellon University, and was able to transmit data about its inventory and the temperature of newly added drinks.[4] In the 1990s, the IoT concept was called the embedded Internet.[5]

IoT's explosive growth in recent years is due to vast improvements in technology in several areas. Let's think back to MyAppoly's new smart fridge, which wouldn't exist without a variety of recent innovations in four primary categories: hardware, networking, big data and data analytics, and cloud technology.

First, Roost was able to purchase or manufacture chips and sensors that were small, cheap, and powerful enough for the fridge's purposes. In the past, it would've been cost-prohibitive or even impossible to manufacture something small enough to fit in a refrigerator yet powerful enough to connect to the Internet. Now, because of advances in hardware technology, we're seeing

---

[3]Dr. JF Fava-Verd and Forster, Sam, "The history of Internet of Things (IoT)," Innovate UK, https://innovateuk.blog.gov.uk/2017/07/03/the-history-of-internet-of-things-iot/.

[4]"The 'Only' Coke Machine on the Internet," https://www.cs.cmu.edu/~coke/history_long.txt.

[5]David Kline, "The Embedded Internet," WIRED, October 1, 1996, https://www.wired.com/1996/10/es-embedded/.

computational components that are getting smaller, cheaper, more energy efficient, and more powerful. Moore's law summarizes these trends in hardware technology improvements.

Moore's law, named after Fairchild Semiconductor and Intel co-founder Gordon Moore, is based on his 1965 prediction that the number of transistors on a computer chip would double every year. What does that mean? Briefly, computers speak in binary, meaning that they only understand information that's written as a series of 0s and 1s. As we learned in Chapter 3, everything we tell a computer to do via human-readable programming languages eventually gets translated into binary so that the computer can understand and execute our instructions. Transistors are the actual pieces of hardware within the computer that represent the 0s and 1s. The more transistors there are on a computer chip, the more computing power that chip has. Over time, as transistors became smaller and computer chips became more powerful, smaller devices gained greater amounts of computing power. The computing power of a handheld smartphone today is far greater than the computing power of a massive supercomputer 30 years ago. Our Roost fridge has the computing power it needs thanks to these advances in hardware technology.

Second, recent networking improvements have led to faster, cheaper, and more reliable connections to the Internet. Needless to say, more widespread and reliable access to the Internet helps Things stay connected. Fast and reliable Internet connections are crucial for IoT networks that we rely on for our health and safety. For example, if smart locks secure the MyAppoly office, the consequences of an Internet outage could compromise the safety and security of the office building.

Third, the Roost IoT network can successfully make use of massive amounts of data thanks to current big data technology. As we discussed in Chapter 5, big data is high velocity (rapidly updated pieces of data), high volume (large amounts of data), and high variety (different types of data). The data generated by IoT networks certainly qualifies as big data and necessitates the use of newer big data technologies. Our Roost fridge generates a massive amount of data about milk weight alone—imagine the sheer volume of data our fridge will have once we stock it with the rest of our groceries and combine data from all Roost fridges in the country! Advances in data mining and statistics, as well as the machine learning advances discussed in Chapter 15, contribute to Roost's ability to glean useful insights from massive amounts of data.

Lastly, advances in cloud technology make it possible for IoT networks to share data and communicate. The IoT platform itself exists because multiple devices can communicate with a central piece of software on the cloud. IoT devices can store massive amounts of data in the cloud that would otherwise be impossible to store directly on the devices themselves. These servers can cheaply store large amounts of data due to the improvements in hardware we discussed above. Powerful servers in the cloud can perform computationally intensive analytics that aren't possible to do on smaller devices.

IoT Applications

The possibilities that IoT brings are endless, and the field is almost certain to grow in ways we can't predict. Some of the many potentially exciting IoT areas today include:

- *Enterprise*: Smart, connected networks have the potential to improve efficiency and productivity in areas such as retail, agriculture, and energy, among others. Consider a sensor network that collects data on farm soil moisture and allows a smart water sprinkler system to improve efficiency and reduce waste in the watering process.

- *Smart cities*: Smart traffic lights, for example, could use historic data to optimize traffic control and decrease congestion.

- *Self-driving cars and connected cars*: Self-driving cars work by using cameras and a variety of sensors, such as radar and LiDAR, to sense and navigate their surroundings. In a future IoT world, self-driving cars could share data with each other to inform each other about traffic accidents and road changes or even share data with traffic lights and construction zones to plan better routes and minimize traffic, among many other possibilities.

- *Wearable technology*: Wearable technology has added potential in the field of healthcare. Wearable devices such as smartwatches and personal fitness trackers could collect unprecedented amounts of very personal, health-related data such as heart rate and hours of sleep. Caregivers could have direct and constant access to patient health information, enabling better healthcare delivery.

- *The smart home*: Everything from light bulbs to doorbells could connect to the Internet and each other.

- *Consumer robots*: Robotics is a complex field that requires a full book to explain in depth. However, there are a few products that we'll quickly discuss here. A robot is defined as a machine that is capable of carrying out a complex series of actions automatically. Some robots can automatically vacuum your floors, some can patrol your house for security, and some can fetch objects. When robots connect to the Internet, their capabilities only increase. For example, a fleet of robotic vacuum cleaners communicating with each other could work together to coordinate vacuuming a large space most efficiently.

Advantages and Caveats

It's clear that IoT technology is extending its reach into many aspects of our lives, including our businesses, homes, and health. The Internet of Things can not only make our lives more convenient but can also make processes and interactions more intelligent, efficient, and accessible. It is estimated that by 2025, the Internet of Things will drive several trillions of dollars of yearly economic impact.[6]

The value that IoT networks can bring includes:

- *Better insights and more effective actions*: As we discussed earlier, MyAppoly employees are benefitting from the insights and actions that their new smart fridge provides.

- *Improved product development*: Roost might notice that after a few months of use, milk weight sensors are slower to share data with the IoT platform. This insight then allows Roost to intelligently put resources into developing a better milk sensor for the next version of the smart fridge, rather than merely making guesses about where their product might need improvement. The data collected about how users and devices are performing allows IoT device manufacturers to create better products more efficiently.

- *Direct economic gains*: Consider a fleet of smart wind turbines that is communicating with each other as an IoT network. During a strong wind storm, Turbine 10 starts to putter out and realizes that it's about to go down. It sends information about its upcoming outage to its surrounding wind turbines, which in turn use the data to adjust their settings dynamically to compensate for Turbine 10's lost time. They also alert smart buildings in the city with an estimate of their temporarily decreased energy supply, so that buildings can conserve energy accordingly. In this way, the economic losses from the outage are minimized. Rather than having to wait for a human to notice that a single turbine is down and adjust the rest of the settings accordingly, a smart network of turbines and buildings can work together to handle the outage quickly.

---

[6]The Internet of Things: Mapping the Value Beyond the Hype. McKinsey Global Institute. 2015.

- *Remote monitoring and control of devices*: Since devices are sending data over the Internet, they can be monitored from afar. Technicians may not have to be present physically at the wind farm in the example above in order to inspect the turbines.

- *Increased machine autonomy*: In our turbine example, the communication between the turbines allows for a human technician to monitor the fleet as a whole, rather than each turbine individually. The human is now free to focus on the overall performance and health of an entire fleet, rather than on the performance and health of multiple individual units.

Of course, the growth of IoT presents new challenges we will need to address. As this growth continues, there are some things we should keep in mind and be cautious about, including:

- *Data security*: Every attack that we learned about in Chapter 12 can target IoT devices. IoT networks increase the number of ways hackers can gain access to data and devices, so protection becomes an even more significant challenge. If a home network is hacked, an attacker could gain control of the smart locks on the home and remotely unlock doors at any time. If a network of healthcare devices, such as a network of patient wearable devices and hospital devices is compromised, an attacker could gain access to personal health information or even control patient medications and treatments directly. A compromised self-driving car could lead to hijacked traffic or car accidents designed by an attacker. It is imperative that we prioritize cybersecurity.

- *Data privacy*: As IoT networks become more commonplace, manufacturers will have access to huge amounts of very personal user data, ranging from healthcare data to conversations with smart speakers to logs of people who have visited someone's home. These data are all private and potentially sensitive, and it's up to manufacturers and the government to develop protocol around (1) how and for how long data are stored, (2) if data are de-identified when aggregated for analysis, and (3) the types of data users can elect to share or not share with other companies or devices. Without comprehensive data privacy practices in place, users are unlikely to adopt IoT networks as part of their day-to-day lives.

- *Interoperability and standards*: IoT devices are manufactured by a variety of companies, using a variety of technologies. How can we anticipate which devices will need to communicate with one another and set standards accordingly?

- *Infrastructure*: Recall that the Internet relies on an infrastructure of physical hardware that exists throughout the world. The growth of IoT devices will strain our current networking infrastructure.

Conclusion

Many technologies discussed in this book come together to create IoT technology, including networking, databases, the cloud, developing at scale, machine learning, and cybersecurity. The Internet of Things is an ever-expanding network of smart devices that are always talking to each other, sharing information, and responding to new knowledge and insights. Though there's a lot we could gain from IoT networks, if the Internet of Things is to grow and become a standard part of our day-to-day lives, we must be extremely cautious about how we protect user data and secure devices.

Artificial Intelligence

Special thanks to Jay Harshadbhai Patel for contributing this chapter.

It is a bright sunny morning and you are getting ready for work. Before heading out, you open your inbox and start reading emails. The inbox has automatically filtered out spam emails into a separate tab so you can ignore them. While replying to an email, you make a grammatical mistake but are promptly corrected by the email service. As it's time to leave for work, your smartphone pops up a notification saying that it will take 24 minutes to reach the office based on current traffic data. You step out of the house, and a self-driving car is ready to pick you up. You enjoy a cup of coffee in the back seat as the car drives you to work. At the office, you "ask" your smartphone "What meetings do I have today?" The phone reads out the details of your meetings. You head to your 10 AM meeting and are well on your way to a productive working day.

In a few years, this anecdote will be commonplace. Artificial intelligence (AI) technologies will make it easier for us to perform day-to-day activities in our personal and professional lives. AI technologies appear to think and to analyze like humans do, but are really just computers crunching large amounts of data to make decisions. Therefore, whereas humans display *natural* intelligence, we say that computers display *artificial* intelligence. In this chapter, we focus on machine learning (ML), a subset of AI that can learn from experience and improve its intelligence continually. Before ML technologies, we had to remove spam email from our inbox manually, but modern email products like Gmail use ML to learn what emails to flag as spam based on the email's content or our past actions. As we use these products more and more, they record the emails, links, and buttons we click and become more accurate. All references to AI throughout this chapter should be read in the context of a self-improving ML system.

Key Aspects of Artificial Intelligence

Every AI technology is built to solve a *problem*, or a task with well-defined inputs and outputs. *Inputs* are data that teach AI to solve the problem, and *outputs* are the solutions. AI's *model*, or brain, is the set of mathematical functions that create outputs from the inputs. The model *learns*, or adapts and improves its behavior, based on the data that it's given. Consider Spotify, a popular music streaming application that also recommends new songs to users. Here, the *problem* is to make song recommendations most likely to please a listener. The inputs are a user's song listening history, trending songs, preferences of similar users, and other useful information that allows it to infer what the user might enjoy listening to. The outputs are the songs that Spotify recommends to their users. The model is the core software component that understands your listening preferences. We can see Spotify learning as its song recommendations improve over time—for example, it might initially recommend top 40 songs since those are popular across the country, but as it receives more information about our past Disney song listening history, it will start recommending more movie soundtracks.

We will dive into each of these four aspects of AI next—the problem, the data, the model, and the learning.

Problems

We use AI to solve problems. The word *problem* is used very broadly here, encompassing everything from fundamental human needs like access to food, shelter, and water to luxuries like having a personal assistant to handle paperwork for you.

There are two main categories of problems that AI is useful for. The first includes problems that are already solvable by humans but where AI can provide a better and more efficient solution. For example, humans have been driving automobiles for decades, yet automated cars have the potential to dramatically reduce the number of road accidents and make possible new ecosystems of car sharing that reduce car ownership and harmful gas emissions. Prominent companies like Google, Uber, and Tesla are pumping a lot of money and talent into building AI to drive cars today. Building this technology requires sensing the environment, identifying vehicles, pedestrians, and obstacles in real time, and acting on these data to navigate the roads.

Another category of problems includes those that are difficult, if not impossible, for humans to solve; here we expect AI to open the doors of innovation. For example, while your closest friends could recommend songs that you might like, they are limited by their own experiences with music. If we could instead browse through all the music out there in the world and recommend those that complement your tastes, you might discover new and exciting music that would otherwise have been inaccessible.

Outputs from AI are usually termed *predictions* to denote educated guesses. Prediction here does not refer to statements about the future. If it returns five new song recommendations to a user, we say it predicted the songs. In other words, it used past data and listening history to determine that the user will enjoy the returned songs the most out of all the songs that it knows of. More generally, we call this a *classification* problem, where the AI attempts to categorize inputs into one or more predefined classes. Here, it predicts "user will like" (encoded as 1) or "user will not like" (encoded as 0) for every song in the knowledge base and returns some of the class 1 songs. There are just two classes (1 and 0), and every song must fall into one of these. The number of classes depends on the problem—more specific predictions require more fine-grained classes. We could train AI to predict the range from "will strongly like" to "neutral" to "will strongly dislike," which will require three, five, or even seven classes. *Regression* problems, in contrast, have continuous, non-categorical outputs. The predictions do not fit into one class or another but instead fall somewhere on a continuum. For example, based on historical monthly (first day of the month) data of housing prices in an area, we can train AI to draw a smooth line that captures the average trend of ups and downs over the years and hence predict future prices.

Data

AI relies on large amounts of correct, varied, and balanced inputs to solve the problem correctly. Since it adapts and improves its behavior based on data, incorrect data means it will learn to predict the wrong classes or values. Housing prices mistakenly shifted by a month will lead to predictions offset by a month. Worse still, corrupted data such as wrong prices will teach AI the wrong things entirely. Even if data is correct but narrowly sourced, it will not be representative of the problem and will lead to poorly performing AI. For example, historical housing prices from the United States should not be used to train AI to predict prices in India since the markets are so different.

Furthermore, the AI model needs to see enough varieties of housing types in data, like single-family homes, duplexes, and mansions so that it is not taken by surprise when asked to predict the price of an obscure house. Lastly, data must have enough quantities of each house type to avoid the majority type from having an overbearing influence on what AI learns.

Computers store all information as numbers, or more precisely bits. Therefore, we must transform data into numbers before AI can learn from it. Thankfully, standard data types like images, text, and audio are already represented as numbers in computers. In practice, we convert data into condensed numerical representations called *features*, which capture the most relevant parts. Let us say you had access to information about all houses sold in the United States over the last hundred years, including details about the houses and their monthly price variations. To build an AI that predicts the price of a house in the future, we might focus our attention on the purchase date, square footage, location, and occupancy (i.e., the features) for each house, since they are most likely to affect prices. We leave out wall paint color and previous tenants' age since we find them irrelevant in determining the price. Combining the most relevant features into a list of numbers gives us a *feature vector*, which is the exact format of input expected by AI.

But how do we know that date, square footage, location, and occupancy are the most impactful factors on housing prices? Maybe wall paint color *is* essential to market price. There are so many things that affect market prices that we might be oversimplifying by manually choosing a few features based on intuition, thus training our AI to learn the wrong patterns. A better way is to automatically discover the most important features through mathematical techniques. We could first plot price (y-axis) vs. square footage (x-axis) for all houses; if the square footage has a substantial impact on price, we should notice a diagonal line indicating that price changes significantly as square footage changes. Then, we can pick this to be one of the features. If the relationship is a horizontal line instead, square footage doesn't impact prices, so we should exclude it from the features.

We could also use simple counting techniques to capture high-level aspects of data and encode them as features. For example, in spam email classification, we could count the top ten most frequently occurring words in spam emails, do the same for all non-spam emails, and remove highly common words such as "a" and "the." Then, AI will learn to distinguish spam from non-spam by checking words in a new email against these word sets and observing whether it should be classified as spam. Google's Word2Vec (short for "word to vector") technique even transforms individual words into mathematical representations that capture their real-life meaning. These features capture semantic aspects of data; for example, the words "mother" and "father" are closely related, whereas "food" and "computer" are unrelated. Features are food for an AI; the stronger their relationship with desired outputs, the more accurate the AI will be (assuming the AI can pick up on those relationships, which is the subject of the "Models" section next).

Taking a step back from computing features, we need to ensure that we are working with the most appropriate data set to begin with. Let's say we wanted to detect pedestrians crossing the road. We could use a microphone to pick up footsteps on the road. However, it would be incredibly difficult to determine how far the person is from the car by the sounds of footsteps alone. A better kind of data here would be images of the road, where a person would be clearly visible. Even better, we might use an infrared image to capture heat signals. Humans emit a lot more heat than the road so they would appear bright red on the image and be easier to detect.

Models

The model is responsible for producing correct predictions given inputs. Input data goes in one end (say a patient's symptoms), and a prediction (flu/no flu) comes out of the other. If the model observes multiple highly related symptoms, it predicts a "yes flu." If it observes a lack of fever, it predicts a "no flu" (since the flu is almost always accompanied by fever). In the middle are hidden mathematical functions that encode real-world knowledge such as the fact that headaches and fever always accompany the flu, while leg pain is an entirely unrelated thing. They are essentially formulas that relate variables with coefficients, such as $y = mx$ where y is severity of flu (the higher, the worse) and x is strength of headache (again the higher, the worse). The variable m determines how x and y are related; if it equals one, then as the headache worsens, the flu worsens to a similar extent. If m equals two instead, the flu doubly worsens, implying a symptom that is very strongly related to flu severity. The example demonstrates that real-world relationships can be expressed as mathematical formulas with variables. Note that these formulas merely indicate correlation between two values, meaning that a headache and severity of flu appear to change together, not that one causes the other to change. Models trained to capture these correlational relationships in formulas are at the core of AI.

Table 15-1 shows the different kinds of correlational relationships between symptoms and severity of flu.

Table 15-1. Examples of Linear and Nonlinear Relationships

Type	Relationship	Example Symptom	Explanation
Linear	Positive	Headache	The worse the flu, the stronger the headache.
	Neutral	Leg pain	The severity of the flu has no effect on leg pain.
	Negative	High mental focus	The worse the flu, the harder it is to focus (i.e., lower mental focus).
Nonlinear	Unclear	Sneezing	The severity of the flu is associated with sneezing in a complicated way.

Linear relationships are simple associations where a corresponding change in one value accompanies an increase or decrease in another value. The $y = mx$ example discussed earlier is a simple linear relationship between y and x. Nonlinear relationships are less straightforward. With old age comes weakened vision, but at a different amount and at a different pace for each person. While one person may have perfect vision even at 80, another may develop nearsightedness at 20. We know that there is some relation between age and eyesight; it is just hard to explain or express. Other real-world examples are how a company's quarterly earnings report affects its stock price the next day or how a new government policy affects unemployment rates in the country. There are no apparent formulas that allow us to compute these effects.

That said, there must be *some* formula hidden from our sights, possibly containing hundreds or thousands of variables, which captures these uniquely complex relationships. A neural network is the most effective type of model for these scenarios. It is built with multiple layers of simple mathematical functions (called neurons) stacked together to produce an output at the end. The stacked functions can go as deep as over 500 layers, giving rise to the term deep learning. You have probably heard this buzzword being thrown around a lot in the media, often touted as a magical algorithm that performs superhuman feats like beating Lee Sedol, one of the best Go players in the world. Neural networks are unique in their ability to capture any linear and nonlinear relationship, allowing them to pick up subtle nuances in data that are otherwise hard to spot. Other types of models include decision trees which are tree-like structures that choose the most informative features to learn, and support vector machines (SVM) which attempt to separate the strongest positive and negative features as far apart as possible.

Learning

Now that we have a defined problem, relevant data, and a model, we can feed data into the model so that it learns. *Learning* is the process by which a model develops knowledge about the problem. When we feed data into a model, it goes through a few data points at a time and adjusts the parameters of its underlying mathematical functions to reflect new information gained from data. This adjusting process is governed by *hyperparameters* (or higher-level configurations) such as how much the model should change its parameters per set of data points, also known as the learning rate. The higher the rate, the more influence each new data point has on the model's parameters and the faster it makes progress. Over time, the parameters' values would ideally get good enough that the model produces accurate predictions, whereby they stop changing and are said to have converged to their final values. In practice, models often do not converge if they fail to capture relationships in data sufficiently. Hence the learning process requires human monitoring at regular intervals to determine if the process is going well, that is, whether the parameter updates are heading in the direction of improving or worsening predictions. It requires multiple passes through the data to converge, taking hours or days for a model depending on the difficulty of the problem. When viewed from the perspective of a human, we call this process training the model. Data being used to train the model is training data.

The three most common types of learning processes are supervised, unsupervised, and semi-supervised learning. Supervised learning relies on human-provided desired outputs called *labels* for each training data point. If the problem is to predict whether a patient has cancer, supervised learning requires not just the symptoms of a patient but also a label "has cancer" or "does not have cancer." The model continually verifies and refines its knowledge by predicting cancer/no cancer for a patient and confirming it with the given label. If it predicts correctly, confidence goes up, whereas if it predicts incorrectly, it's time to change some values in the mathematical functions that underlie the model. This iterative process of checking and updating adds to the model's knowledge over time. At the point of convergence, we stop training the model, validate it by giving it new patients' information, and count the number of mistakes it makes. The model has not seen these patients before, so it is a test of how well it understands the problem. If it is good enough, we might integrate the model into a real product and sell to consumers. If it is not, we would continue training for a longer time with some changes to the underlying mathematical functions or structure, such as adding more layers to a neural network or tuning some hyperparameters. Sometimes we might collect more labeled training data to give the model more information about the problem.

Unfortunately, labels require costly manual curation. For the problem of detecting pedestrians in road images, it is relatively easy to collect thousands of images, but much harder to go through each one manually and assign a yes/no for the presence of a pedestrian. Unsupervised learning develops knowledge from training data in the absence of labels. The iterative process of verifying against true labels in supervised learning cannot work here. Instead, this process attempts to uncover relationships within features, such as identifying similar data points and placing them into groups. The problems that unsupervised learning can solve have more to do with gaining insight into data rather than trying to predict an output. The K-means algorithm is one such learning process that iteratively discovers clusters in data. For example, if the data contains customer information such as age and hometown for a bookstore, grouping them helps us better understand the kinds of people that frequent the store.

In many cases, only some of the data have labels. Can we still make use of them to do learning? Semi-supervised learning attempts to tackle this case, mostly by inferring the labels on all other data points depending on their similarity to the small set of labeled ones. The more similar, the more likely they are to have the same label. For example, if two patients had very similar symptoms, but we only knew the label for one of them, we might make an educated guess that the other also has the same label. Once we have all the labels, we can apply the previously discussed supervised learning approach to learn and predict outputs.

Now we will see how these four key components of AI, problem, data, model, and learning, come together in a small real-life case study.

Case Study: Apple Face ID

To unlock the latest iPhone, you are able to look at the screen for a split second merely. This technology uses a neural network that has been trained to detect your face.[1] Here, the problem involves classification mainly—determining whether the face looking at the screen right now is the actual owner of the iPhone. To perform this task accurately, the model is first trained to understand what human faces tend to look like with training data that includes three-dimensional depth maps and two-dimensional infrared images of over a billion real-world faces. That's right, Apple sought consent from lots of different people to save pictures of their faces in various orientations! From this data, the model tries to learn common facial patterns such as the fact that most people have two eyes, one nose, and one mouth. This is an

---

[1]Apple FaceID Security Guide https://images.apple.com/business/docs/FaceID_Security_Guide.pdf

unsupervised problem setting since it is more about uncovering relationships between parts of the face than trying to match a face against human-curated labels. In fact, there are no labels in this context. Diversity in training data is crucial to allow a model to pick up on the nuanced relationship between aspects of facial structure, that is, different people have different eye colors or distance between the eyes. These relationships are captured as formulas which are highly adaptable even if you wear glasses, hats, or other things that change the appearance of the face. When you first buy an iPhone, you are asked to calibrate it with your own face so that the model can further tweak its formulas from the training data to adapt to (or learn from) your particular face; since you will be the sole user unlocking the iPhone, it needs to recognize just you and no one else. In the future, unlocking the phone is as simple as matching a new face against this stored representation. Over time, the model continually learns to account for changes in facial structure so that the phone still unlocks if you grow a beard or long hair.

Ethics of AI

AI models inadvertently capture and exacerbate discriminatory biases present in society. In 2015, Google Photos tagged multiple photos of two black people as "gorillas," which led to an uproar from many users on Twitter and forced Google to issue a public apology for the mistake. While the exact cause was not made public, we can speculate that the training data had an insufficient number of examples of black people for the model to learn them accurately, or that the model was not able to capture differences between black humans and gorillas. It is also possible that training data were inaccurately labeled, that is, pictures of gorillas were actually labeled as black people, so naturally the model learned from that. Google might have been able to prevent this mistake by using a more diverse and accurate dataset and a better model. They could have performed more thorough testing by feeding a representative variety of images into the photo tagger and making sure the outputs were sensible. As a potential creator of new AI technology, you should be mindful to anticipate biases in the data and actively remove them while training a model. It is also crucial for the team that is building AI to be composed of diverse individuals, including but not limited to different genders, ethnicities, and races. This makes it all the more likely that inconsistencies and biases in data are accounted for since each member of the team brings a unique lens of society with them.

AI in Practice

AI is a fast-moving field where new tools and techniques are being developed all the time, so we might expect this list to evolve rapidly. Many of these tools and techniques are publicly available with documentation and tips on getting started; with just some programming experience, you can train a simple machine learning model on your personal laptop in less than an hour. This makes it incredibly simple to embed AI into your own product since you do not have to reinvent the wheel.

Table 15-2 shows some of the tools people use in 2019 to actually incorporate AI into their products.

Table 15-2. Tools to Integrate AI into Products

Tool	Creator	Use
Tensorflow	Google	Build and train complex neural networks in Python
PyTorch	Facebook	
Caffe	Berkeley AI Research	
Caffe2 (merged into PyTorch as of March 2018)	Facebook	
Scikit-learn	Open source	Build and train machine learning models in Python and some simple neural networks
Numpy	Open source	Store and manipulate large amounts of data and feature vectors. One of the most compatible data formats for feeding into Tensorflow and PyTorch
ConvNetJS	Andrej Karpathy (previous PhD at Stanford)	Build and train neural networks in a web browser using JavaScript

Conclusion

AI is a powerful tool to build products with predictive capabilities that impact millions of people. From suggesting songs to detecting cancer, it is widely applicable to a range of scenarios where large amounts of data are available. Nowadays, it is easy for someone with programming experience and access to a fast computer to train a model on their own dataset without much mathematical background. This means we will see more exciting applications of AI in the future. And we must be aware of potential biases in the datasets that determine what the models learn since ultimately they will affect.

The Blockchain

Special thanks to Vojta Drmota for contributing this chapter.

Blockchain technology, sometimes referred to as *distributed ledger technology*, is one of the latest innovations in technology. It has applications in a whole host of fields including finance, law, economics, mathematics, philosophy, and computer science. This chapter covers the fundamentals.

Introducing Blockchain

One of the pillars of modern technology is how data is stored. Due to the vast amounts of available data, it is important to store it in a way that will make it easily accessible later on. You might choose to store your personal budget in a spreadsheet with rows and columns, which makes it easy for you to navigate visually. Many companies with immense quantities of data store it in relational databases, which are similar to spreadsheets but are also easily navigable programmatically. A *blockchain* is just another way in which data can be stored.

Data stored in a blockchain is packaged into *blocks,* which are linked together to form a linear chain: the blockchain. The blockchain can be thought of as a book, where the book in its entirety represents the blockchain and each individual page represents an individual block. Blocks in the blockchain are numbered the same way as pages in a book are, starting from the first block, known as the *genesis block*, and continuing to the last. The page number, which here represents the position of the block in the blockchain, is known

© Vinay Trivedi 2019
V. Trivedi, *How to Speak Tech*, https://doi.org/10.1007/978-1-4842-4324-4_16

as the *block height*. Pages in a book have uniform sizes and can thus contain a predetermined maximum number of words. Likewise, blocks in the blockchain have a uniform data storage capacity and can store a predetermined maximum amount of data.

There are two things, broadly speaking, that readers of a book take for granted when flipping through its pages. Firstly, readers assume that every page that is supposed to be in the book is, indeed, in the book. If a student were to buy a history textbook in which the chapter, or even a couple pages, on World War II was missing, the student would be robbed of key information. In other words, we expect identical editions of books to contain identical information. This expectation applies to the blockchain as well. Each copy of a specific blockchain needs to contain all relevant blocks—and therefore all relevant data—in order to be *valid*. Otherwise, two parties sharing a blockchain to store data will have contradicting data sources. Secondly, we take it for granted that the order of the pages in a book is correct. A student attempting to understand the timeline of World War II, for instance, would be misguided if the pages on the conclusion of the war were, without logical reason, before the pages on the causes of the war. Since data often follows a logical order, it needs to be stored logically. This applies to the blocks in a blockchain too: a block that contains data on you spending Bitcoins cannot come before a block that contains data of you receiving Bitcoins—that would mean you're spending Bitcoins before you earned them. Therefore, the order of blocks needs to remain intact for the blockchain to be *valid*.

What does it mean for a blockchain to be *valid*? The answer involves the very reason why the blockchain came about as a way of storing data in the first place. Storing data on a blockchain doesn't allow a company to store more data than it otherwise could store using a traditional database. Nor does storing data on a blockchain allow a company to navigate the data more efficiently than traditional databases do. So, why would someone store data on a blockchain? The advantage of storing data on the blockchain comes from its unique ability to allow multiple, independent parties to add data to a shared data storage. For this reason, the blockchain is sometimes referred to as a *distributed database*. For instance, ten private companies maintaining vaccination records could store them on a blockchain in order to share the data, ensuring that no client gets the same vaccination twice. A blockchain is *valid* when at least a majority of the companies involved agree that the blockchain is storing the data appropriately. You'll find out exactly what this means later in the chapter.

Why can't these companies just share a central database stored on a server owned by one of the companies? They surely can, and this was the primary solution before the blockchain was invented. Yet, the central database solution has the major disadvantage of having a single point of failure. What if the company in charge of the servers shuts them down unexpectedly? What if the company doesn't properly back up the database and suddenly loses all the data? What if the company decides it wants to shut another company out of the system and unilaterally revokes their access? In other words, the central database solution requires all parties to trust a single, central authority. If the data is stored on the blockchain, on the other hand, there is no need for a single, central authority to maintain a server with the central database. Instead, each of the ten companies maintains a copy of the data and independently verifies the legitimacy of any new vaccination record that is added. This lack of a central entity that needs to be trusted for data to be collectively stored and retrieved is known as *decentralization*. Each individual or institution participating in a blockchain is known as a *node*.

What Does a Block Look Like?

Blocks are the fundamental pillars of the blockchain. They store all information and are linked together to form the blockchain. Each block in the blockchain is represented by a *digital fingerprint*, which is unique to the block and can be used to unambiguously identify the block in the blockchain. A digital fingerprint works very similarly to a human fingerprint. Humans can be unambiguously identified with their fingerprints, meaning that if someone leaves their fingerprints at a crime scene, investigators can use the fingerprints to match them against fingerprints previously collected in a database. However, the person cannot be identified from the fingerprints alone: a fingerprint database needs to contain a mapping between fingerprints and other forms of identification, such as people's names and photos of their faces. Digital fingerprints of blocks work in the same way: nothing about the data in a block can be deduced from the block's digital fingerprint, but a block can be unambiguously identified in the blockchain using its digital fingerprint. This is useful when, for instance, checking whether one copy of the blockchain is missing any blocks.

Other than its own digital fingerprint and any included data, each block also contains the digital fingerprint of the block immediately before it. A blockchain is a linear data structure, which means that each block is *appended* to one block and has at most one block that is, in turn, appended to it. The blocks are chained together by storing references to the digital fingerprints of the block to which they are appended, which also makes it easy to maintain and validate the correct order of the blocks.

A digital fingerprint is simply a 64-character-long hexadecimal number. Below is an example of a digital fingerprint:

0f978112ca1bbdcafac231b39a23dc4da786eff8147c4e72b9807785afee48bb

The *hexadecimal numeral system*, unlike the decimal system humans conventionally use, contains digits with 16 values (0–f) rather than 10 values (0–9). The hexadecimal system is used because it allows for more numbers to be represented using fewer characters than the decimal system. This is good for data storage efficiency. For instance, the largest possible number that can be represented in the hexadecimal system using only three digits is *fff*. In the decimal system, this number would be 4095, which is significantly larger than 999, the largest possible three-digit number in the decimal system.

The Science of Mining

Mining is perhaps the most elusive concept in blockchain technology. Let's begin by understanding the need that mining fills. Recall the example of ten private companies using a blockchain to share a database of their clients' vaccination records. There was no central authority, no single company in charge of maintaining the database and ensuring that all the data in it is correctly stored and accurate. Instead, each of the ten companies individually vetted all incoming data and maintained a local copy of the blockchain. How do the companies agree on what data to add and what to throw away if there is no central authority to stipulate this? The ten companies involved in the blockchain need to come to a *consensus*, or agreement, on what data to add. Mining is a process by which this consensus is reached between multiple nodes. It was pioneered in the Bitcoin cryptocurrency, the first widely used application of blockchain technology. The process of mining in Bitcoin will be explained later in this chapter, but a core part of the mining process in all blockchains is the creation of blocks' digital fingerprints.

The creation of digital fingerprints involves cryptography, which relies heavily on mathematics. A digital fingerprint is created by running the block data through a *hash function*. A hash function takes any input data, usually text, applies a mathematical algorithm to the data, and outputs a value that uniquely represents the input data. Digital fingerprints for Bitcoin blocks are created by running the block data through the SHA256 algorithm, a hash function designed by the US National Security Agency (NSA). SHA256 takes any arbitrarily sized data as input, manipulates that data, and outputs a 64-character-long hexadecimal number, known as the *hash*. Regardless of the size of the inputted data, the output is always 64 characters long. Any change in the input data will produce a new, unique hash, and the same hash cannot, in practice, be produced by two different inputs. This makes it very easy to detect if a block's data has been tampered with. Thus, the hash of a block's data, colloquially referred to as the block's *digital fingerprint*, is used as a block identifier. Table 16-1 shows a few examples of digital fingerprints generated from arbitrary input data:

Table 16-1. Digital Fingerprint Examples

Input Data	Digital Fingerprint: SHA256 Hash	Notes
a	ca978112ca1bbdcafac231b39a23dc4da786 eff8147c4e72b9807785afee48bb	One character produces a 64-character-long hash.
Blockchain	ef7797e13d3a75526946a3bcf00daec9fc9c9 c4d51ddc7cc5df888f74dd434d1	An entire word produces an equally long hash.
Blockchains	99cf6497afaa87b8ce79a4a5f4ca90a579773 d6770650f0819179309ed846190	One small change in the word produces a completely new hash.
The favorite number of the fox was 44.	f70d31700c3c122331538f9b389a10217 e0bd1cc0694b67a5c7b4f02c17b6198	Regardless of the input data's length, the hash is 64 characters long.

You might have heard that mining is a *computationally intensive process*. However, generating a hash from any input data, such as the above, takes a fraction of a second and can be done without using too much of a computer's resources. Generating a hash, in other words, is *not* a computationally intensive process in and of itself. What makes mining computationally intensive is that a specific type of hash needs to be generated from the input data in order to create a valid digital fingerprint for a block. In the Bitcoin blockchain, for instance, this entails generating a hash from the block data whose numerical value is smaller than a given *target value*. More information about the mining process can be found in the Bitcoin section of this chapter.

The term *miner* is used to denote the parties involved in the mining process. However, the term "miner" does little to clarify the purpose of their job. The primary purpose of miners is to maintain the validity of a blockchain in lieu of a central authority. As stated above, miners undergo a computationally intensive process in order to find the digital fingerprints of new blocks. This process is a substitute for a central authority that stipulates which blocks are valid and contain data that should be appended to the blockchain. Exactly how this process achieves this goal differs from blockchain to blockchain, but the most popular method is found in the Bitcoin blockchain, which is detailed later in this chapter. Miners aren't individuals sitting at their desks iterating through hashes: a miner is a piece of software running on a (very powerful) computer that automatically iterates through millions, billions, even trillions of hashes per second. These computers are owned by independent individuals or companies who usually receive a reward for running their mining software. In Bitcoin, for instance, miners create and receive new Bitcoins every time they find a valid digital fingerprint for a new block. As a result, they are generating new coins and making money by performing their job as miners. However, not all blockchains let miners generate new coins when new blocks are found, and some blockchains have no rewards for miners at all. In such cases, miners are instead referred to as *validators*.

Immutability of the Blockchain

The blockchain is sometimes referred to as a *ledger* because of the way data modifications are recorded. Think of how you would change a value, such as the bank account balance of a user, in a spreadsheet or traditional database. You would most probably overwrite the cell or data field and replace the old value with the new value. This method of modifying data is not possible on the blockchain. Instead of overwriting the data in a block, you would append a new block to the blockchain that modifies that data. As a result, just like a ledger, the blockchain keeps track of every data change ever made. This integral property of the blockchain is known as *immutability*. Note that this is different from saying that a blockchain cannot grow: on the contrary, blockchains are immutable precisely because the only way they permit changes to the data they hold is through the addition of new blocks, that is, through growth. The Bitcoin blockchain grows at an approximate rate of one block every 10 minutes, but once a block is validated and appended to the blockchain, it should, in theory, remain unchanged forever. Figure 16-1 shows an example of the process.

If we consider Figure 16-1 to be the current state of a blockchain that has a block height of 3, then Alice, currently, has 7 points. Note that instead of simply updating the first block with Alice's current point total, we added a new block each time her point total changed. Hence, the blockchain serves as a ledger containing a transaction history, or the history of data manipulation. Due to its immutable nature, the blockchain is sometimes referred to as an *immutable ledger*.

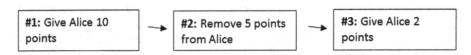

Figure 16-1. Data modification on the blockchain

Both the data in the blocks as well as the order of the blocks, known as the *block order*, are immutable. In the example above, the order might seem irrelevant as addition is commutative, but blockchains are usually governed by complex rules that require immutable ordering. Consider, for instance, that a rule governing the blockchain above is that no user is allowed to have more than 10 points. If the order were somehow disrupted, and block #2 and block #3 swapped places, Alice would have 12 points in the now-second block, which would render that block erroneous. This erroneous block, known as an *invalid block*, would render the entire blockchain invalid. Consequently, both the block order and block data should, in theory, not be changed in a blockchain.

Decentralization

As mentioned earlier, the blockchain is an appealing way of storing data because it allows multiple, independent parties to share a database in the absence of a central authority. This means that there is no master copy of the data, no central authority that decides what data can or should be added, and no single point of failure. This type of network is known as a *decentralized network* because it lacks a central hub. This type of network is at odds with how most of society is structured. Take currencies, such as the US dollar, for example: it is backed by a central government, controlled by a central bank, and transacted via banks who act as centralized intermediaries. When we use the US dollar, we trust these centralized institutions to act with integrity and fulfill the duties they promise to fulfill. As soon as a central bank fails in its duties, the currency collapses. This happened, for instance, when Zimbabwe's monetary policy caused hyperinflation in 2008 and the Zimbabwean dollar lost all of its practical value. On the other hand, a currency that runs on the blockchain, like Bitcoin, has no central government or central bank controlling its functions. Instead, blockchain technology is used to distribute these roles, traditionally centralized, among miners who collectively maintain the network. In other words, blockchain technology allows for a *trustless network* to exist, whereby you don't have to trust anyone while still being certain of the integrity of the system.

While there are many benefits associated with decentralized networks that are built on top of blockchain technology, the implications of an absence of a central authority need to be kept in mind. If you accidentally send your Bitcoins to the wrong person or feel that you have been scammed, there is no central authority you can turn to in order to have your Bitcoins returned. Likewise, there is no bank that can vouch for the safety of your funds or aid you in major money transfers. A decentralized network removes the safety nets that central authorities provide in centralized networks and leaves individuals to fend for themselves. This is an important consideration to keep in mind when engaging in decentralized networks, especially those that involve high risk. Currently, the most popular decentralized network built on a blockchain is the Bitcoin cryptocurrency.

Bitcoin: The Genesis of Blockchain Technology

Bitcoin is a *cryptocurrency*, a form of digital money, that uses the blockchain to function as decentralized money. Bitcoin, along with most cryptocurrencies, is, therefore, an application of blockchain technology—currently the most popular application. Bitcoin is the oldest cryptocurrency and currently has a market cap of over a billion US dollars. In fact, blockchain technology arose from the creation of Bitcoin: the inventor of Bitcoin, an anonymous individual or group of people publishing under the alias *Satoshi Nakamoto*, invented blockchain technology to serve as the transaction ledger for Bitcoin. Before this invention, digital currencies were either centrally controlled or could be very easily manipulated and destroyed. Only after Bitcoin's creation did people realize the potential of blockchain technology in other domains.

Bitcoin was invented in 2008 when Satoshi Nakamoto published a white paper entitled "Bitcoin: A Peer-to-Peer Electronic Cash System." A year later, in 2009, Nakamoto published the first piece of Bitcoin code. The code was released as *open source*, meaning that anyone could read it and suggest improvements to it, which resulted in a community-driven effort among programmers to constantly improve the Bitcoin code. Nakamoto's publications were made on an online forum for peer-to-peer technology enthusiasts. *Peer-to-peer technology* is a type of technology that is intended to run without the need for a centralized authority or server. Fiat currencies, such as the US dollar, are not peer-to-peer networks, because banks act as central authorities and oversee all transactions, the storage, and the creation of money. As discussed earlier, Bitcoin is a peer-to-peer network without a central authority that dictates the validity of transactions or maintains the power to shut the system down. No individual or institution can manipulate or exert control over Bitcoin. As such, Bitcoin is a *decentralized network*.

The key invention that Nakamoto made in Bitcoin was the *proof-of-work algorithm*, which lays the foundation for how Bitcoin as a network achieves consensus or, in other words, agrees upon which transactions are valid and which are not. In fiat currencies, this is not a problem, because banks centrally stipulate validity: if you try to send someone money that you do not have or if you try to spend your money twice, your bank will prevent you from doing so. Therefore, when using a fiat currency, you trust banks as central authorities to make correct decisions about transaction validity. In Bitcoin, however, *trust is decentralized*, meaning you can trust no one in the network yet still be certain that no invalid transaction, such as someone stealing your Bitcoin, will take place. To attain this, Nakamoto invented proof-of-work as a mechanism through which *distributed consensus* is reached, whereby the Bitcoin network agrees on a set of rules that dictates transaction validity. These rules are rooted in mathematics, specifically in cryptography.

To understand Bitcoin's distributed consensus, it is important to understand some basic demographics of the Bitcoin network. The users of the Bitcoin network are loosely divided into two groups. The first, larger group is comprised of the regular users who send and receive Bitcoins; the second, smaller group is comprised of the *miners* whose job is to validate transactions and group them into blocks, which are then appended to the Bitcoin blockchain. In theory, anyone can become a miner and thereby contribute to Bitcoin's distributed consensus.

Miners have tangible incentives to continue performing their job of mining new blocks. Each time a miner successfully mines a new block, they reap the *block reward*. The block reward is a predetermined amount of Bitcoins that are created in the new block and sent directly to the miner. This special type of transaction is known as the *coinbase transaction*. Currently, 12.5 Bitcoins are created in every new block and sent directly to the miner. The block reward halves every 210,000 blocks. On top of the block reward, the miner of the block receives all *transaction fees* that are contained within the block. Every Bitcoin transaction contains a transaction fee, which is designated and paid by the user sending Bitcoins. The higher a designated transaction fee, the quicker the transaction will be confirmed on the Bitcoin blockchain because of the greater incentive for miners to include it in their blocks.

Next is a walkthrough of the process of how Bitcoin as a currency can work without a central bank or government.

How Bitcoin Works

The process of how Bitcoin as a currency works can be abstracted into five steps.

Step 1: A New Transaction Is Created

Alice wants to send 5 Bitcoins to Bob for a new car she purchased. Normally, if Alice used a fiat currency such as the US dollar, her bank would transfer the funds to Bob's bank. Alice's bank would remove the money from her account, and the money would appear in Bob's bank account. With Bitcoin, however, there are no banks, and so the process of transferring Bitcoins is very different. Let's walk through how it happens.

To begin, each Bitcoin user has a *private key* which needs to remain personal and secret. The private key is used to sign all transactions made by the user. This is an important part of Bitcoin because, in the absence of banks, it serves as the mechanism through which a user exerts control over their money. Each user uses their private key to verify their ownership of their Bitcoins when sending them. In other words, this private key is used to unlock the user's Bitcoins. If the private key is lost or stolen, the user loses access to his or her Bitcoins.

The *private key* is a random 64-character-long hexadecimal number (note that this is different from the digital fingerprint of a block). There is no central database that stores all existing private keys. If you are concerned that someone might create the same private key as the one you are already using—don't worry! There are 2^{256} private keys available, which is more than the number of atoms in our universe. The probability of someone randomly generating a private key that is identical to yours is lower than that of you walking outside right now and being killed by a falling piano.

A *public key* is generated from the private key using elliptic curve cryptography, an advanced approach to cryptography. This public key is then hashed and encoded to create a unique *Bitcoin address* for each user. The Bitcoin address is meant to be public as it is the address to which you receive Bitcoins. In our example, Bob would have to give Alice his Bitcoin address (Table 16-2).

Table 16-2. Example of a Bitcoin Private Key and Public Address

Private key	e9873d79c6d87dc0fb6a5778633389f4453213303da61f20bd67fc233aa33262
Bitcoin address	1BoatSLRHtKNngkdXEeobR76b53LETtpyT

Alice verifies ownership of her 5 Bitcoins by using her private key to create a *cryptographic signature* of the Bitcoin transaction. This cryptographic signature serves as proof that Alice did indeed send 5 Bitcoins that were hers to Bob. Any miner can verify that Alice did indeed sign the transaction by using Alice's public key to verify the transaction data. Hence, Alice never needs to share her private key.

Once Alice signs the transaction, her *Bitcoin client* (the piece of software Alice uses to handle Bitcoin payments) broadcasts the Bitcoin transaction, containing information that it is meant for Bob, to the Bitcoin miners through a *gossip protocol*.

Step 2: Miners Verify the Transaction

Each miner verifies this transaction upon receiving it by, among other things, checking whether Alice indeed has 5 Bitcoins to spend. This is done by traversing across the Bitcoin blockchain and subtracting all the Bitcoins that Alice has spent from those that she has received. Recall that the blockchain is an immutable ledger of all transactions that have ever taken place. This means that if a miner starts with the first block and reads data from every single block until the latest block, he or she is able to calculate Alice's balance.

Step 3: The Transaction Is Waiting to Be Mined

Once the miners validate the transaction, it is stored in each miner's *memory pool*, which is a local data storage that holds all valid transactions that are waiting to be packaged into blocks and added to the Bitcoin blockchain. Note that when a transaction is in the memory pool, it is not in the Bitcoin blockchain, that is to say, it is not yet a confirmed Bitcoin transaction. The transactions sitting in the memory pool are waiting to be mined into blocks and appended to the blockchain.

Step 4: The Transaction Is Packaged into a Block, Mined, and Appended to the Blockchain

How are transactions taken from the memory pool, mined into blocks, and appended to the blockchain? This involves some complexity because we need to make sure that each miner does not add a different block to the blockchain and that the transactions are valid. In other words, all the miners need to verify transactions and agree on a single block that they all add and do so without a central authority telling them what to do. One way to make miners agree is to have them play a random game that only one miner will win and that requires them verifying the transactions as part of the game. When that miner wins, the transactions are verified and that miner earns the right to add his/her block to the blockchain. Everyone else must agree to do the same. We call this game "the Bitcoin proof-of-work algorithm."

As an example of such a game, imagine a game where Player A thinks of a number between 1 and 10. Player B's task is to guess the number, but she can only guess one number per try. Assuming that Player A thought of her number randomly, there is no deterministic methodology that Player B can employ to figure out the number. Simply, she has to guess. This type of guessing is known as *brute-force* guessing. If Player C, then, joins the game but with two guesses per try instead of just one, she is statistically expected to guess the number quicker than Player B. This is because brute-force guessing can be made more efficient with an increased frequency of guesses.

Miners engage in a similar brute-force guessing game when attempting to find the digital fingerprint of a block that is valid. To begin the process, the miner selects an arbitrary amount of transactions from the memory pool. The miner then appends some required data, such as the current *time stamp*, and runs everything through the SHA256 hashing algorithm. The resulting hash is the *block hash*, also referred to as the block's digital fingerprint. This process of selecting transactions and hashing them to create a block is known as *mining*. How does the miner know if he or she found a valid block hash? If the numerical value of the block hash is lower than the current *target* value, the block is deemed valid and can be appended to the Bitcoin blockchain. See Table 16-3. An intuitive way of comparing the hexadecimal values of SHA256 hashes to the target is by looking at the leading zeros: the more zeros at the start of the hash, the lower its value.

Table 16-3. Judging the Validity of a Block Hash Based on the Target Value

Example target value	00000000000000000000365a1700000000000000000000000000 000000000000000
Invalid block hash	000f6497afaa87b8ce79a4a5f4ca90a579773d6770650f0819 179309ed846190
Valid block hash	00000000000000000000000000000000a4a5f4ca90a579773 d6770650f0819179309

Statistically, more often than not, the miner will find a hash that is greater than the target and therefore generate an invalid block that cannot be appended to the Bitcoin blockchain. To generate a new block hash, the miner can either change the permutation of the transactions in the block, select a new set of transactions from the memory pool or—and this is the preferred option— add a random number to the data in the block, known as a *nonce*. Recall that the smallest change in input data will produce a new hash, so the miner can iterate through different nonce values to produce radically different hashes.

Miners iterate at a *hash rate* of trillions of hashes per second, often using specialized hardware, in a competition to be the one that finds the next valid block hash. The faster a miner can iterate through hashes, just like Player C had more guesses per try, the more likely he or she is to find a valid digital hash. This computationally intensive process is known as proof-of-work because once a miner finds a hash, it serves as proof of the work he put in to mine the next block.

Once a miner finds a valid block hash, the block is broadcasted to the Bitcoin network, and other miners verify and append it to their version of the Bitcoin blockchain.

Step 5: The Other Miners Verify the Block Containing the Transaction

Although mining a block is a computationally intensive process, verifying that a block is valid is a simple job. Once the other miners receive the mined block, all they have to do to validate it is to run it through the SHA256 algorithm once. If the block hash value is below the target, the miners accept the block as valid and append it to their local copy of the blockchain. The block has now been *confirmed*. If it sounds like this is a really complicated process, it is.

If you want to create a process that does not require a central authority figure, you need to get creative and, in this case, you need to get technical. There is a tradeoff here though—in exchange for creating a decentralized system, the "game" that miners need to play slows things down. Transactions cannot be processed as quickly as some would want, which is why some are starting to create their own versions of the blockchain, as discussed later in the chapter."

Once a block is appended to the blockchain, it is almost impossible to remove or change the block. The deeper a block is in the blockchain, measured by the number of blocks that come after it, the harder it is to remove or change that block. Recall that blocks contain links to previous blocks; thus, in order to re-mine a block (a technical way of describing the process of changing a block's contents) that is 5 blocks deep, an attacker would not only have to re-mine that block but also the 4 blocks that come after it and any new block that has been added to the blockchain during this time. It is, in practice, impossible for a single miner to have the electrical output to produce a hash rate high enough to perform such an attack.

Therefore, the deeper a block is in the blockchain, the more immutable it is and the more Alice and Bob can be certain that their transaction has been securely approved. The deeper a block is in the blockchain, the more *confirmations* it is said to have. Hence, a block that is 5 blocks deep is said to have five confirmations. A transaction is considered securely final after *six confirmations*.

This five-step process demonstrates the simple rules that are followed by thousands of independent miners to asynchronously achieve consensus about the state of the Bitcoin blockchain. The *state of the Bitcoin blockchain* is a technical way of saying the number of Bitcoins that each user holds. The process is asynchronous because there is never a point in time when an election or ruling takes place that stipulates a consensus on the state of the blockchain. Instead, the proof-of-work algorithm is intended to act as a lottery of which miner gets to produce the next block and consequently extend the blockchain. As such, no central authority, such as a bank, needs to be trusted in this decentralized network.

The Price of Bitcoin

The *price of Bitcoin* is dictated by supply and demand. At any given point in time, the value of Bitcoin is the value at which it was traded in the latest transaction. For example, if Alice sends half of a Bitcoin to Bob for $500, then the price of one Bitcoin at that given point in time is $1,000. Bitcoins can be bought on online *exchanges*, such as Coinbase or Bitstamp. Most exchanges determine the price of Bitcoin by aggregating the prices of the latest transactions.

Since there is no central bank that controls the value of Bitcoin through monetary policy, Bitcoin has thus far suffered from high *price volatility*. Many critics of Bitcoin argue that price instability prevents mainstream adoption of Bitcoin as a medium of exchange and renders Bitcoin a speculative instrument instead. Proponents of Bitcoin argue that price volatility will decrease as Bitcoin usage rises. If more merchants begin accepting Bitcoin payments, fewer users will seek to convert their Bitcoins back to fiat currency. This will lower speculation levels and begin stabilizing the price. Currently, however, Bitcoin and many other cryptocurrencies face high price instability.

Storing and Losing Bitcoins

Users store Bitcoins in pieces of software known as *wallets*. Despite the name, wallets do not actually store your Bitcoins; rather, they scan the Bitcoin blockchain to calculate how many Bitcoins you have access to with your private keys. The wallet software also takes care of generating private keys and creating corresponding public keys and addresses for Bitcoin users.

There are two ways in which your Bitcoins can be *stolen*. The first method involves the thief getting access to your private keys. Once the thief gains access to the private keys, they can use them to transfer all Bitcoins associated with the private keys to themselves. The second method involves a hacker hacking into the servers of an exchange where Bitcoins are traded. Most exchanges store the private keys to the Bitcoins that are currently being traded in a database. If the hacker gains access to this database, he or she can transfer any Bitcoins associated with the private keys to himself or herself.

The Bitcoin network itself is vulnerable to attacks, primarily to attacks launched by malicious miners. The most prominent of such attacks is known as the *51% attack*. Recall that consensus in the Bitcoin network is distributed, whereby simple rules are followed by many independent miners to produce new blocks. If a miner decides to disobey the rules and begins mining otherwise invalid blocks, the Bitcoin network will simply disregard him or her, unless the dissenting miner comprises a *majority*. A majority, in this case, is measured in hash rate, meaning that if a miner, or a group of colluding miners, controls more than 51% of the power to produce new blocks, they can dictate what blocks get produced. In such a scenario, the malicious miners can undo transactions and spend their Bitcoins more than once. As a result, the decentralization of the Bitcoin network is undermined. Due to the vast amount of miners, it is very difficult to conduct a 51% attack on Bitcoin because of the sheer amount of energy, primarily electricity, required as input to conduct such an attack. An attacker would need the electrical output of all of Austria to conduct such an attack. To date, there has not been a known instance of a 51% attack on the Bitcoin network.

What has happened, however, is that a minority of miners have disagreed with some of the rules that govern Bitcoin and decided to change them. When a group of miners disagrees with the rules that are followed in the distributed consensus of the Bitcoin network, they may choose to no longer participate in that network as miners. This is known as a *fork* because the dissenting miners split away from the Bitcoin blockchain and begin working on their own blockchain, which entails the creation of a new cryptocurrency. The most popular Bitcoin fork happened on August 1, 2017, when the cryptocurrency Bitcoin Cash was created. The miners who split away from the Bitcoin network to create Bitcoin Cash disagreed, among other things, with the Bitcoin consensus rule that stipulates that valid blocks have a maximum size of 1 MB. The miners wanted to increase the block size in order to increase the number of transactions that can be validated in one block. The block size of Bitcoin Cash blocks was increased to 8 MB.

The Times Ahead

Many skillful developers constantly work on improving the Bitcoin code. Currently, most efforts are geared toward improving the speed and efficiency of Bitcoin transactions. A new technology known as the *lightning network* is being developed, which will allow very small transactions to be sent instantaneously across the network. There is also a lot of work being put into spreading awareness and adoption of Bitcoin. While Bitcoin's price volatility is rampant, it is lower than that of the Venezuelan bolivar, the national currency of Venezuela. This has resulted in individuals converting their savings into Bitcoin as it is easier to access than the US dollar. Moreover, several merchants in Venezuela have begun accepting payments in Bitcoin. The more central institutions fail in maintaining national currencies, the more can we expect Bitcoin adoption to rise.

The World of Blockchains

Cryptocurrencies have thus far been the most popular blockchain application, which is mainly due to the fact that blockchains are a great way of keeping track of digital asset ownership. Since the creation of Bitcoin, countless new cryptocurrencies have spawned, some more legitimate than others. Each new cryptocurrency exists on its own blockchain with its own network of users and miners. While new cryptocurrencies bring varying degrees of new innovations to the table, most are, at least conceptually, based on Bitcoin. For instance, the cryptocurrency Monero was created to allow private transactions by obfuscating transaction data stored in the blockchain. Dash, another cryptocurrency, was created to reduce Bitcoin's transaction costs and increase transaction speeds. Each of these cryptocurrencies exists on separate blockchains and requires specialized software to be used. There are currently thousands of cryptocurrencies; however, not all cryptocurrencies actually provide what they promise. Most cryptocurrencies don't gain enough traction to survive or are outed as frauds.

Ethereum

The most notable post-Bitcoin cryptocurrency is Ether, which is built on the *Ethereum* blockchain. The reason why, unlike Bitcoin, the Ethereum blockchain is referred to as an ecosystem is because Ethereum provides more than just a cryptocurrency. In Ethereum, users can upload pieces of code that will run on the blockchain. This is useful because it allows any programmable job— not just monetary transactions—to be recorded, validated, and executed in a decentralized manner. The same way Bitcoin erased the central authority of financial institutions in currency, Ethereum can erase the central authority of any institution overseeing digital assets.

Let's look at an example. Imagine the job of a contract lawyer in making and enforcing a legal agreement between two parties. Among the most pertinent tasks of the lawyer is to oversee the integrity of the contract, ensure the contract goes unchanged after signing, and execute the contract agreement once the criteria are met. Sound like the lawyer is a central authority and a single point of failure? He is indeed. If the lawyer is corrupted or makes a mistake, the contract can be manipulated. In other words, both parties involved in signing the contract trust the lawyer to perform his job veraciously: it is a centralized network.

If the contract were instead written in code and uploaded onto the Ethereum blockchain, it would be enforced programmatically by each miner on the Ethereum network. This doesn't mean that the miners would warp into the real world and force a party to abide by the contract. They do, however, provide a provable record of what, according to the contract, should be done. Hence, the same way Bitcoin intends to replace banks as financial intermediaries, so does Ethereum intend to replace contract-enforcing bodies as intermediaries in agreements. These programs that run on the Ethereum blockchain are known as *smart contracts*. When multiple smart contracts interact to form a more complex program, it is referred to as a *decentralized application* or *dApp*. Decentralized applications don't have to be fancy: they can be a simple set of rules such as "breed a new kitty when 0.5 Ether is received." This is actually an example of a rule in one of the most popular Ethereum dApps: a game called Cryptokitties.

Initial Coin Offerings

Initial Coin Offerings (ICOs) have been the cause of a lot of hype and controversy in the world of blockchains. Think of an ICO as something similar to an unregulated Initial Public Offering (IPO). Instead of selling shares, however, a company that wants to do an ICO will sell *tokens*. There are two types of tokens that can be sold in an ICO: *utility tokens* and *security tokens*. Utility tokens are the more popular type as they fall outside of most legal regulation. When you purchase a utility token, you aren't buying a share in the company but a discounted coin that can be used to interact with the company's dApp when it launches. On the Ethereum platform, for instance, Ether is used to pay miners to mine transactions in smart contracts, so Ether can be regarded as both a utility token and a cryptocurrency. Owning a security token, on the other hand, entitles you to certain ownership rights of the company. These tokens are less popular in ICOs as they are regulated by legislative bodies. ICOs are wholly unregulated, meaning a company can determine how many tokens it wants to sell, autonomously set the price of a token, and sell it to virtually anyone.

In mid-2014, a year before the Ethereum platform launched, 60 million Ether coins were sold to investors in a presale that would become the first of many ICOs launched on the Ethereum platform. However, ICOs can take place on any blockchain platform. In 2017 alone, companies raised a whopping $3.25 billion through ICOs. As an investor in ICOs, it is of crucial importance to do your due diligence before purchasing any tokens. Due to the lack of regulation, there is nothing stopping a company from disappearing and not meeting anything on its roadmap after it makes money from selling its tokens in an ICO. There have been many ICO scams in the past, and it remains a difficult job to safely navigate the ICO landscape.

Blockchain Beyond Cryptocurrencies

Other popular applications of blockchain technology exist in *decentralized supply chains* and *decentralized property transfers*. Both of these applications make use of the blockchain's ability to provide individuals with ownership over digital assets or digital representations of physical assets. For instance, a smart contract can be used to handle real estate transactions with minimal involvement of a broker while the blockchain keeps track of who owns what property. Similarly, a supply chain can be made more efficient by representing a good on the blockchain and requiring each supplier to cryptographically sign when and where they processed the good. Thus, it is possible to efficiently track the movement of a good through a supply chain all the way to its source and identify any faulty suppliers. There also exist ambitions to fully decentralize certain supply chains, which would entail a retailer purchasing a good directly from the producer of that good. The energy sector is a particularly salient industry for this type of decentralization as, for instance, homeowners with solar panels on their roofs could sell excess energy to their neighbors. Some of these blockchains, if operated by private companies who exert control over who gets to be a node or a miner, are known as *private blockchains* because participation in the distributed consensus is limited to a private party. While private blockchains have a limited ability to achieve full decentralization, they can nonetheless be used to distribute authority and trust among multiple actors.

Conclusion

Blockchain technology allows us to create decentralized and trustless networks that connect people and companies in novel ways. Blockchains can be used to create networks without central authorities and increase the security of supply chains while cutting administration costs. The direction of blockchain technology is perhaps most eloquently embodied in a message left by Satoshi Nakamoto in the first Bitcoin block:

The Times 03/Jan/2009 Chancellor on Brink of Second Bailout of Banks

The message, citing the headline of a 2009 issue of *The Times*, is a sharp criticism of the blind trust we sometimes place in centralized institutions. When these institutions fail in their roles, it is often the everyday users who pay the price, just like citizens end up paying the bailout of a failing bank. Blockchain technology allows us to build fully functioning networks without the risks associated with central authorities.

Virtual and Augmented Reality

Special thanks to Khoi Le for contributing this chapter.

Ding. The elevator doors open on a magnificent skyline... and then you look down. You're standing on the edge of a building, a hundred stories off the ground. At your feet, a wooden plank extends six feet over the ledge in front of you. "Walk," you hear a voice say. Against your better judgment, you inch out slowly, taking step over cautious step. As you reach the edge, you feel your stomach drop, and you almost lose your balance. However, you stabilize as the wind whistles around you. "Step off the plank," the voice says. At this point, most people refuse to step off the plank, paralyzed by fear. Why would you want to fall to your death? However, you know that you are currently in virtual reality, seeing all of this through a headset strapped onto your face. You're in a classroom, so there's no way you'd get hurt. Of course you'd step off the plank. But your feet won't move because you're afraid.

V. Trivedi, *How to Speak Tech*, https://doi.org/10.1007/978-1-4842-4324-4_17

The plank experience demonstrates the power of *immersive media*, which tricks your brain into believing that the digital, something produced by a computer, is real. *Virtual reality (VR)* creates entirely digital worlds that seem real. A closely related field is *augmented reality (AR)*, which creates believably real digital objects that look as if they are part of the world. VR and AR comprise the bookends of what many call "immersive media."

Why is it called "immersive" media? Imagine walking through the ruins of Pompeii in Italy. The buildings were destroyed thousands of years ago. Luckily, you're wearing an augmented reality headset. In addition to seeing the ruins, there are holograms, digital replicas of the buildings exactly how they used to look based on historical research, overlaid on top of the ruins. Because the experience is spatial and the digital structures are superimposed on the real world in the correct context of the ruins, it augments your reality. Immersive media powerfully surrounds you with digital objects that appear real in order to give you new experiences.

Traditional media are "pancake" media, flat experiences on paper or screens. Immersive media are spatial—they are experienced in three-dimensional space. You can think of immersive media as the spectrum depicted in Figure 17-1, from reality, where we only perceive real objects in a real environment, to virtual reality, where we only perceive virtual objects in a virtual environment.

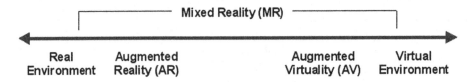

Figure 17-1. The reality-virtuality continuum, from a 1994 paper by Paul Milgram

Mixed reality (MR), a "mix" of virtuality and reality, encompasses the rest of the spectrum. For example, Pokemon (virtual objects) on the sidewalk (real world) counts as "mixed reality." Conversely, a real chair (real object) perceived as being in a virtual living room (virtual world) also counts as "mixed reality." Augmented reality counts as mixed reality since it shows virtual overlays, or objects overlaid on the real world. Imagine looking at someone and seeing their name hovering above their head.

Many people also refer to immersive media with the umbrella term *XR*, where X stands as a placeholder for "virtual," "augmented," "mixed," or anything else people come up with, and "R" stands for "reality." XR unlocks the potential of *experience*. Just as the Internet is a boon to the cataloging, distribution, and democratization of *information*, XR promises to do the same for *experiences*. Students learning about the ocean can go underwater; surgeons, quarterbacks, and police officers can train in realistic scenarios; and anyone can gain empathy by experiencing a snapshot of life as a Syrian refugee, a woman in rural India, or a homeless person. VR/AR is a set of new media through which we can experience connection, stories, entertainment, and education.

In this chapter, we'll explore immersive media through five areas that define VR/AR. First, we'll look at the "New Dimension" that makes spatial experiences so unique. Then, we'll look at "Immersive Content," analyzing the different types of content that exist in VR/AR experiences and exploring content creation processes. After we learn about the content, we'll see how today's "Hardware" allows us to consume that content through tracking. Once we understand tracking, we can discover how avatars and virtual humans represent "You, Virtually" in VR/AR. Finally, we'll look at how people interact with virtual environments in "The Metaverse." These five sections will equip you with everything you need to know about VR/AR.

A New Dimension

To understand how virtual and augmented reality work, we must dig into the crux of immersive media: 3D. This section will first explore stereoscopic vision, which is the way that humans visually perceive the three dimensions. Then we'll look at how computers create three-dimensional scenes and how the limitations of the technology affect the VR/AR experience.

Stereoscopic Vision

If you close one eye and lift your finger to your face, then close the other eye instead, it appears as though your finger is at a different position. Each eye sees the world from a slightly different perspective. Your brain combines these two images to give you a unified image with depth information. Depth information informs you how far away an object is, which allows you to perceive the three dimensions. The process by which your eyes see two images and your brain combines them is called *stereoscopic vision*. Stereoscopic vision, or "stereo" for short, is what makes classic 3D movies seem like they pop out at you from a flat screen. The red and blue glasses show you slightly different perspectives on the same image, giving you depth information. Similarly, VR/AR headsets draw a slightly different image to each of your eyes to provide you with the perception of depth and space in a process called *rendering*.

Rendering takes place in the *graphics processing unit*, also known as a *GPU*, which is a specialized processor in computers that focuses on drawing things onto the screen. GPUs can only draw on the screen so fast, so as you move your head around, it's scrambling to render the scene for you in real time.

Although you cannot consciously tell, your eyes are incredibly quick and can catch images that display for durations as short as one two-hundredth of a second. That means that a GPU needs to be able to keep up with your visual system; if a GPU cannot render new views to the screen fast enough, your visual system becomes confused. When the GPU can't keep up, your eyes see old images that are lagging behind. If you were previously looking around, but now are not moving, your eyes are still seeing moving images. Conversely, if you were previously not moving, but now are looking around, your eyes are seeing stationary images. So your vestibular system—responsible for balance and motion—tells your brain something different from your visual system. Much like seasickness, this conflict between your vestibular system and visual system can lead to sickness. The VR/AR version of nausea is known as *simulator sickness*.

Depending on the complexity of the images, GPUs draw at a certain speed, known as *frames per second*, or FPS. A GPU running at 60 FPS means that the GPU can draw 60 different views every second. A GPU's FPS directly affects the VR/AR experience. To ensure that users do not get sick, developers typically aim for 60 FPS on smaller GPUs in mobile devices. 60 FPS is minimally comfortable but is the fastest rate at which most small GPUs can draw. On computers where the GPUs are larger, developers have the luxury to aim for at least 90 FPS to be sure that all users have a comfortable VR/AR experience.

In a VR/AR headset, known as a *head-mounted display* (HMD), each eye is shown a different image to enable stereoscopic vision and allow users to perceive depth and space. While a typical phone or computer only has to display one image at a time, an HMD has to show two different images simultaneously. Because of this dual display, the GPU must draw two versions of the scene every second, halving its efficiency. Thus, for you to have a bearable VR/AR experience, the GPU needs to be twice as powerful. Due to advancements in the AI industry, which relies on GPUs for intense calculations, only recently have sufficiently powerful GPUs become cheap enough to open the gate to consumer VR/AR. Even so, when building VR/AR content, one must take the necessary precautions to optimize the rendering pipeline to avoid simulator sickness.

Immersive Content

The two primary forms of VR/AR content are computer graphics and video. Computer graphics are visuals created by the computer. Video is just a recording shown in a 180-degree hemisphere or 360-degree sphere, in which you can look around. We'll explore how to make each and what each modality is useful for.

Computer Graphics Content

In computer graphics, "3D models" are described by a few need-to-know terms. *Meshes* are like wire sculptures that serve as the frame for a 3D model. Meshes are composed of faces known as *polygons*. Each 3D model has a "polygon count," or *poly-count*, which shows the number of polygons in the model. Models with high poly-counts are consequently more detailed, making GPUs work harder, which may lead to rendering at a lower FPS. These *FPS drops* cause lag and, consequently, simulator sickness. To combat this phenomenon, many VR/AR artists use a *low-poly* style, seen in Figure 17-2, which uses a very low polygon count. This makes it easier for the GPU to render and also contributes a unique look. The earliest computer VR setups had to use low-poly environments made of wireframe cubes, which looked like they had no surfaces.

Figure 17-2. Google Daydream Home uses a low-poly art style[1]

To control what a model's surfaces look like, artists use materials. Materials wrap around the wire sculpture, like a skin that covers the sculpture's frame. Materials are typically composed of one or more textures, which dictate the pattern and color on the skin, and one or more shaders, which control how the material will show up on screen. For example, a reflective shader would display light emanating off of the model.

How to Make 3D Models

Now that you know that the mesh, polygons, and materials comprise a 3D model, we can explore two common ways of creating 3D models.

---

[1]©2018 Google LLC, used with permission. Google and the Google logo are registered trademarks of Google LLC.

The first method we'll look at is manually creating 3D models. 3D artists use very complex 3D modeling software to arrange each polygon in 3D space by hand. They also create the materials, textures, and shaders that will alter the appearance of the mesh. This is an incredibly labor-intensive process that requires expertise and an eye for detail. In order to optimize models for VR/AR, 3D artists need to be especially cognizant of the poly-count.

3D Models by Photogrammetry

Another method for creating 3D models is *photogrammetry*. You take photos of the desired subject, say a statue, from multiple different angles. Then, the photogrammetry software can infer the shape of the object. Photogrammetry is useful for getting detailed models of real-world objects that would take too long or be unfeasible to model with the desired level of detail.

A Brief Talk About Animation

Once you have made the model, you might want to *animate* it (a static puppy wouldn't be much fun!). Most animation software uses *keyframes*, which are certain important positions that the model will move between. For example, if you have a waving arm, you can make a keyframe of the arm to be leaning to the right. Then, in the second keyframe, you can set the arm to be leaning to the left. The animation software figures out how to get the arm from the first keyframe (leaning to the right) to the second keyframe (leaning to the left), so the arm will go from right to left, appearing to wave. Animating is extremely complicated for complex models with intricate systems of movement.

3D Models with Animation by Volumetric Capture

An alternative to animation is *volumetric capture*, which is similar to a dynamic version of photogrammetry. Volumetric capture typically uses over one hundred cameras simultaneously taking photos, so it can handle moving objects and people in order to create a sort of "3D video." Volumetric capture is great for recording predefined sequences such as a dance routine.

Immersive Video

Immersive video content is another manifestation of VR/AR content. Video in VR/AR is a lot like normal video, but you can look around. Imagine a panorama photo that is moving all around you. These are typically shot with 360 cameras or special 360 camera rigs, which take several images and *stitch* them together. *Stitching* is the process of finding the overlaps between the images and lining them up so that they look like one image. Immersive video is useful for journalism, since it can bring people into another part of the world or into someone else's habitat. It is less useful for exploration, as users cannot move around, only look.

Now that we have our VR/AR content, let's check out how the hardware enables us to look and move around.

The Hardware

Virtual and augmented reality are heavily reliant on the hardware experience. In addition to an HMD, VR/AR uses a tracking system and may have additional hardware components such as hand controllers. The most important thing to understand is *tracking*, which is how the hardware knows where you are and where you're looking. We'll explore different types of tracking and how different VR/AR systems vary in their tracking.

Tracking

To correctly render a VR/AR view, the computer needs to know both where you are looking and where you are located in relation to the virtual environment. These two data points define the two major levels of tracking capability: *three degrees of freedom* and *six degrees of freedom*.

Degrees of Freedom (DOF)

If you think about a light dimmer that slides up and down and only up and down, that dimmer has one degree of freedom. It can only move vertically. Similarly, a world globe has one degree of freedom: it can rotate around one axis. Now, imagine a game like Super Mario Bros. Mario can run left and right but also jump up and down. That's two degrees of freedom. A degree of freedom is simply any value in a system that can vary on a range.

VR/AR has three degrees of freedom: *pitch*, *yaw*, and *roll*. Look up at the ceiling and down at the ground. Pitch is the rotation around the x-axis, or forward-backward rotation. Look left and right. That's yaw, or rotation around the y-axis. Tilt your head left and right, moving your ears toward your shoulders. Rotation around the z-axis is known as roll. These three degrees of freedom are the only ones registered in 3DOF VR/AR, so 3DOF VR/AR only changes the view based on where you are looking. Even if you move forward, it will look like you're standing in place. But you can turn your head all you want. Thus, most 3DOF experiences don't want you to move around but rather just look around. You can think about it as "swivel chair XR."

With *6DOF*, you unlock three additional degrees of freedom: motion along the x-, y-, and z-axes. In other words, you can move forward, backward, left, and right in addition to ducking or jumping up and down. 6DOF VR/AR changes the view based on where you are looking and where you are located. We'll call this "walk-around XR."

More specifically, 3DOF and 6DOF describe the capabilities of specific hardware components. For example, a 3DOF hand controller can point at different angles like a laser pointer, but it cannot track the position of your hand. On the flip side, a 6DOF hand controller will accurately track both the orientation and the position of your hand.

So how does the computer know where you are looking and where you are located? It is easy to know where you are looking. *Gyroscopes* are small instruments inside phones and VR/AR headsets that can tell which way a device is oriented. However, finding out where you are located is extremely difficult.

Imagine driving a spaceship in pitch black outer space. How would you know where you are or if you are even moving? Driving in the dark is what a computer does when it is figuring out where a device is in 3D space in the real world. Computers cannot see, so they cannot tell where a device's position is or if it is moving. So we're in the dark. Now if there were lights in the distance, you could tell where you were in relation to those lights. Alternatively, if you could suddenly see all the asteroids and planets in front of you, you could see how you are moving in relation to those. If there are points of reference for a computer, then it can know where it is. There are two ways for a computer to have points of reference. One way is for external, outside points of reference to send a signal into the computer about its location; this is known as *outside-in tracking*. The other way is for the processes inside the computer to look out for points of reference in the environment; this is known as *inside-out tracking*.

Outside-In vs. Inside-Out

Found in the Oculus Rift and HTC Vive, *outside-in tracking* is the most common tracking method right now. External sensors live on your desk or your walls and act as points of reference, like the lights in the distance from our space example. They do this by sending signals to the sensors inside the headset. From this data, the processor can triangulate the position of the headset over time. Outside-in tracking is extremely accurate but requires external sensors to be plugged in and set up. Outside-in tracking limits you to the space where your sensors are.

Setting up sensors can be tedious, and moving your VR setup (computer, sensors, HMD) is much work. Moreover, AR headsets would be extremely pointless to use in a confined space, since the power of AR is in providing information about the world. The solution is to put the tracking system inside the headset, known as *inside-out tracking*. There are no external sensors since the HMD uses internal cameras and depth sensors to see the environment and calculate where the HMD is in relation to everything. The spaceship can now see the asteroids and figure out its relative position. This process is called *SLAM*, which stands for Simultaneous Localization and Mapping, and is also used by robots and self-driving cars to know where they are in the real world. Without external sensors, everything is more convenient and portable. Inside-out tracking allows for fewer wires and less setup, making VR more accessible. Moreover, inside-out is extremely important for AR HMDs, as it will enable them to move freely around the world.

With augmented reality, a prevalent method of tracking is *markered tracking*. As the name suggests, markered tracking uses a physical marker to help with tracking. A marker is a specific image, typically a QR code, that is printed on a card. However, a marker could be a logo or even a painting.

The camera on a device can recognize this pattern and render a 3D model on top of that image. Marker-based tracking feels like a 6DOF experience because an image on a card can move in all six degrees of freedom, so the 3D model rendered on top of the card can as well. However, the algorithms don't actually know the position of the device, just where the tracked marker is. Marker-based tracking is very accurate, but ultimately lacks the contextual knowledge about the environment that will make future AR very powerful. Most AR tracking nowadays will implement some form of SLAM, which we discussed earlier, since it is possible without markers and external sensors. However, marker-based tracking creates a smooth user experience for certain constrained use cases like the classroom.

VR Systems

Now that we know how tracking works and what types of tracking exist, let's look specifically at the different kinds of virtual reality systems.

Smartphone VR

The most popular form of virtual reality is *smartphone-powered virtual reality*. Smartphone-powered VR works by rendering two images on your phone screen that you can look at through magnifying lenses, typically inside an HMD container that holds your phone. Smartphone VR is most commonly used to watch 360 videos. Smartphone VR utilizes 3DOF tracking through the phone gyroscope, which, you'll remember, measures orientation. More recently, 3DOF hand controllers have also become available for smartphone VR. The popular names in this space are the Google Cardboard and Daydream as well as the Samsung GearVR. Although fairly accessible, smartphone VR does not compare to the experience of PC VR.

PC VR

PC VR requires a Windows computer with a hefty GPU, attached to a VR HMD. There are two types of PC VR: tethered (HMD is attached to PC with cable) and untethered (no cables). Most VR game developers are making PC VR games. Playing games on PC VR is an incredible experience. The graphics are incredible because the GPUs are larger and the tracking is more precise due to computational power and mandatory external sensors. VR arcades are exclusively full of PC VR systems, and most universities use PC VR systems for research and development. Furthermore, PC VR is currently the only VR system that has 6DOF hand controllers. The most popular systems are the Oculus Rift and HTC Vive. However, PC VR is extremely costly, since you have to own a computer with a fancy GPU, a fancy headset, and a large, empty space to use it in. It is also challenging to transport around and requires setup before use. To make VR more accessible, standalone VR emerged.

Standalone VR

The latest release in the VR hardware ecosystem is standalone virtual reality. Standalone VR places all of the necessary hardware components, such as screens, GPU, and processor, inside the HMD, eliminating the need for an external phone or computer. Standalone VR is the most compact, portable, and cost-efficient solution on the market. There are 3DOF and 6DOF tracked standalone HMDs. For example, the Oculus Go has a 3DOF HMD and a 3DOF controller. The HTC Vive Focus has a 6DOF HMD, but its controller is only 3DOF. Later in 2019, the Oculus Quest will be released. It will be the first commercially available consumer 6DOF HMD with 6DOF controllers. Standalone virtual reality is a massive step toward consumer virtual reality since people can now buy one device and immediately experience VR.

AR Systems

Much like VR, AR has several different classes of systems.

6DOF HMDs

These HMDs show 3D graphics that live in the space of the real world. You can move around the objects, and the HMD uses inside-out tracking to know where you are and anchor the digital objects as if they were in your space. The major players here are Microsoft HoloLens, Meta, and Magic Leap.

Heads-Up Display (HUD)

If you know what Iron Man sees, these HMDs are very close to that. They layer information, typically text or basic images, into the world in front of you. However, these aren't rendered into your world as objects—rather, they simply show in front of your eyes. These are useful for industrial workers who need to see information but don't have the hands to hold a phone or tablet. Similarly, skiers or motorcyclists who need information about their speed can wear these types of display HMDs. A popular term for these HMDs is *smartglasses*. A smartglass brand you may have heard of is the Google Glass.

6DOF Mobile AR

Mobile AR, unlike mobile VR, actually refers to mobile phone-based augmented reality. Mobile AR uses your phone's camera to read in the real world and then displays 3D graphics overlaid on a feed that shows up on your phone's screen (Figure 17-3). For example, IKEA has an app that allows you to see virtual pieces of furniture, like a chair, appear in your living room. Mobile AR uses 6DOF markerless tracking, so you can move around the chair as if it were in the real world.

Figure 17-3. Mobile AR on an iPad showing a digital forest[2]

Mobile AR relies on intense *AR-specific SDKs*. Vuforia is a large-scale software suite made for image detection, marker tracking, and other useful needs. Although Vuforia is still powerful and useful for higher-level AR functions, Apple's ARKit and Google's ARCore have recently become powerhouses at the lower level because they are hardware specific and do great markerless 6DOF inside-out tracking using SLAM.

With all of the different hardware systems in the VR/AR landscape, it is easy to get a little lost. However, remember that the main differences are tracking: 3DOF (swivel chair XR) vs. 6DOF (walk-around XR) and outside-in (external sensors) vs. inside-out (SLAM).

You, Virtually

One of the powers of virtual reality is enabling you to be whoever you want to be. You'll also interact with a wealth of virtual characters in VR and AR. A representation of someone in VR/AR is called an *avatar*. Although avatars can be anything, most avatars are humanoid because that accurately maps to the tracking systems that exist, which track the head and hands with six degrees of freedom.

Moreover, people are generally more comfortable portraying themselves as humanoid and seeing other humans with humanoid avatars, known as *virtual humans*. Avatars are incredibly prevalent in virtual and augmented reality because when we enter a virtual space, we need a personal manifestation to show to others in that same virtual space. According to communication and media researchers Blascovich and Bailenson, virtual humans display multiple types of "realism."

Photographic Realism

If you look at yourself in the mirror, a high-quality video, or a photo, what you see has extremely high photographic realism. It seems just like a real person: you. In VR/AR, there are rarely photorealistic avatars due to both graphics and a phenomenon known as *the uncanny valley*. Modern computer graphics capabilities are phenomenal and getting closer and closer to making realistic looking virtual humans. However, in VR/AR headsets, the rendering costs are still high, so most systems cannot afford vivid, realistic details.

Moreover, as virtual humans approach realistic human resemblance, the uncanny valley causes discomfort. On one side of the valley are cartoony virtual humans, while on the other side are completely realistic virtual humans. Humans are typically comfortable on either side of the valley, but right in

---

[2]Photo by Patrick Schneider on Unsplash (https://unsplash.com/photos/87oz2SoV9Ug).

the middle, where the virtual humans almost look like real humans, people can get uncomfortable. An excellent example of the uncanny valley is the creepiness of the characters from *The Polar Express* (2004), which caused a lot of discomfort for viewers. This discomfort stems from the fact that our minds are biologically wired to try and identify people. In the uncanny valley, our brains can't decide whether an image is a real human or not, making us uneasy. As a result, near-photorealistic virtual humans are generally avoided.

Most VR/AR experiences, such as *Rec Room* (2016), use cartoony virtual humans, nonhuman representations, or human representations with helmets or masks. Well, you may ask, what value does VR/AR have if avatars don't look like real people? Photorealism is actually not the most important type of realism. Behavioral realism has a much larger impact on perceptions of realism.

Behavioral Realism

While photorealism measures "does it look real?", *behavioral realism* measures "does it ACT realistically?" Even if an avatar does not look like a real person, you will treat it like one if it behaves like one. Think about talking to Alexa or Siri. You use a lot of the same mannerisms that you would use when talking to a person. VR's main challenge with behavioral realism in avatars is in tracking. Since AR and VR tracking systems are similar in nature, but VR tracking is a little farther along, we will focus on VR-tracked avatars. We will focus on VR-tracked avatars. For most 6DOF PC VR systems, you can only track the six degrees of freedom for the head and hands. However, you may want to show the virtual human's entire body to display a full range of motion. You might want body language to come through, for the elbow to move in a realistic manner, or for the legs to move when your avatar walks around. To get from the head and hands to an entire body, people use inverse kinematics, also known as IK. IK algorithms use the positions and rotations of the head and hands to guess where the elbows and hips are. There are still many behavioral realism challenges such as facial expressions. Making mouths behave correctly during talking, creating realistic eye movements, and having detailed hand behavior are all problems being tackled today.

The Metaverse

Now that we have a representation of ourselves, let's explore the virtual world. Virtual and augmented reality are powerful media because of *presence*, the psychological feeling of "being there." Many people who experience deep presence will remember becoming extremely disoriented after taking off the headset because they are now in their living room instead of on top of a castle fighting enemies. So what are the various elements inside a virtual environment that contribute to presence?

A virtual environment has many components. Because VR/AR is currently primarily an audiovisual experience, the main components discussed below will be spatial audio and the primary visual elements of the environment, including the set dress and skybox.

Spatial Audio

Spatial audio is how sound works in the real world. VR/AR is a spatial medium, so spatial audio is a perfect fit. In an experience with spatial audio, as a virtual ambulance zooms by you, you would hear the siren fade in and out. If you are in VR/AR and put your right ear up to a virtual boombox, the music in your right headphone would increase. Spatial audio helps create ambience and presence since it replicates how humans perceive sound in the real world.

Set Dress

Set dress is a game development term that is carried into the design of virtual environments. Set dress is the small details that make an experience feel real. Set dress could be a virtual coffee mug on your desk in AR or a dying houseplant in the corner of the VR living room you're standing in. All these small items add character and authenticity to a virtual environment.

Skybox

In virtual reality, some virtual environments will take place in an outdoor setting, like a city or a forest. A *skybox* is an image that dictates how the horizon and sky in VR will look. The name sky "box" comes from the fact that a giant cube surrounds you, where each of the cube's six faces has a different photo. The top face usually has an image of what the sky looks like, the four walls show the horizon, and the bottom face has an image of the floor—which most people won't see because virtual environments typically already have floors. Skyboxes aren't always boxes, they can also be giant hemispheres.

Haptics and Interactions

When your body is involved in a VR/AR experience, there is a higher sense of presence. To increase presence through body involvement, creators should build interactions into the environment, complete with *haptics*.

Controllers

The feeling that your hands are there in the virtual environment is known as *hand presence*. In an AR environment, you're able to see your hands in front of you, so hand presence is generally not an issue. In a VR environment, however, the headset blocks your view of your own hands, which might make you feel like your hands aren't even there. Hand controllers in virtual reality add a lot to presence because they give the user both agency and psychological *body transfer*. Body transfer occurs when a person starts to feel that they have ownership over a body that is not their own. Body transfer increases presence because if you own your avatar's body, you can cause things to happen in the virtual environment. You have *agency*. Agency is the amount of control a user has to affect their environment. For example, you might design an experience such that the user can knock objects off a table. Having environmental interactions with haptic feedback in VR increases agency and thus presence.

Haptic Feedback

Haptic feedback refers to forces or vibrations that recreate the sense of touch for a user. The most common haptic feedback you may have experienced is the vibrating sensation when you type on your phone's keyboard. Similarly, in virtual reality, the most common form of haptic feedback comes from a controller that vibrates when you touch a virtual object. Other types of haptic feedback include feeling temperature through a specially programmed heated fan that blows hot air in your face if you are near a virtual fire, feeling physical objects like a wall where a virtual wall is, and feeling resistance through special haptic devices. Having tactile haptic interactions in immersive media drives agency and connection with the experience, as touch can evoke emotions. Imagine the experience of a virtual bullet hitting you or a virtual hug surrounding you.

Moving in VR: Locomotion

Locomotion refers to the act of moving around a virtual environment. Unfortunately, you cannot move more than 10 feet before running into a wall in your room or reaching the end of the cable connecting the computer with the HMD. The virtual world you're exploring is much larger than 10 feet—so how can you explore virtual worlds if you can't normally move around them? There are many methods of locomotion in VR, but people are constantly experimenting with new and creative ways: teleportation, hand-based locomotion, and infinite treadmills. The most common mode of locomotion is *teleportation*. You simply aim your controller at a point on the ground in the distance and then you will appear there.

Teleportation can break presence, however. There are also other locomotion experiments, such as the notion of climbing. Since you only have your head and hands in VR, many people experiment with hand input for locomotion. Climbing involves using the hand controllers to grab the virtual floor or walls and pull yourself around the environment. There are also many hardware experiments for locomotion. Though not required, hardware peripherals can augment a VR experience. One interesting concept is the *infinite treadmill*, which allows you to run in any direction while staying in one spot. Currently, there are several different iterations on locomotion hardware, such as a bowl that you run in while wearing slippery shoes or a treadmill that goes forward, backward, left, and right, and can consequently move "diagonally" at different rates.

Moving in AR: The AR Cloud

Moving in augmented reality is the same as moving in reality since the virtual environment exists in your real-world environment. Here, the issue isn't about how you move around the space but instead about where the space is rendered in relation to you. Imagine that you've put a virtual mug on the very corner of a table. How will your friend walking by with her AR glasses on see the same virtual mug on the very edge of the table? Her AR glasses have to recognize that she is in the same room as you and recognize the exact table that you put the virtual mug on. To have ubiquitous, shared augmented reality all over the world, we need the *AR Cloud*. The AR Cloud is a one-to-one mapping of our world, essentially a digital copy of all the places and spaces on earth. The AR Cloud will have to keep track of all the 3D spatial data and meshes of our world and all the users who use AR and where they are in order to build large-scale multiuser experiences. While this has a long way to go, it will enable a powerful digital layer to exist overlaid onto the real world. In the meantime, large-scale multiuser AR experiences cannot leverage the precise anchoring of 3D models or ubiquitous multiplayer, which limits the possible shared experiences of AR.

The "Metaverse" is continually evolving, and all of the components above are important to creating powerful immersive experiences. If we're to live in a world where the line between digital and real is blurred, let's build a good blend.

Conclusion

The year is 2049. Everyone has contact lenses that allow them to see digital objects rendered everywhere. You can watch 360 videos of your child's birthday or look at 3D models of a new house you want to buy. At stores, augmented reality virtual humans, looking impeccably photorealistic, stand behind the counter, ready to take your order. Your friends can leave secret messages around the world that only you can see. You can beam into work from the comfort of your home or visit family on the other side of the world.

Virtual and augmented reality are here to stay. Although there are many terms and concepts involved with immersive media, this chapter covers all the need-to-know ideas in VR/AR. We explored the notion of three-dimensional experiences, rendered by GPUs and perceived with stereoscopic vision. We compared videos, which represent our world, to 3D computer graphics models that compose new worlds. We learned about the difference between 3DOF swivel chair experiences and 6DOF walk-around experiences, and we covered the various immersive systems that exist today, such as standalone VR/AR. The fourth section covered avatars, examining the relationship between photographic and behavioral realism in virtual human design. Finally, we closed with the environment and interactions of VR/AR, learning about the different aspects that add to presence. You're ready to go out and confidently speak about immersive media and how it will become the next generation of human-computer interaction. Now, you are fully immersed.

Index

A

Access control, 106

Advanced Research Projects
    Agency (ARPA), 2

Advanced Research Projects Agency
    Network (ARPANET), 1

Ajax, 29

Amazon Elastic Compute
    Cloud (EC2), 12

Amazon Web Services (AWS), 12

Android, 114–115

Application programming
    interfaces (API), 46
  advantages, 47
  authentication
    client-side certificates, 52
    message-based, 52
    open API, 51
    SSL endpoint, 52
  disadvantages, 47
  documentation, 50
  JSON, 52
  MyAppoly API
    advantages, 48–49
    disadvantages, 49
  REST, 50
  SOAP, 50
  working principle, 49

The AR cloud, 172

Augmented reality (AR), 158, 166–168

Avatars, 168

B

Back endtechnology. See Programming
    languages

Backups, 108

Berkeley Software Distribution (BSD)
    license, 56

Bitcoin price, 151–152

Blockchain, 139

C

Cascading Style Sheets (CSS), 26

CIA Triad
  availability, 99–100
  confidentiality, 95–97
  integrity, 97–99

Client-driven approach, 67

Cross-platform mobile development, 115

Cross-site request forgery (CSRF), 98

Cross-site scripting (XSS), 98

Cryptocurrency, 142, 146, 153–155

Cryptography, 101, 142

Cybersecurity, 94

D

Database management system (DBMS), 33

Database systems
  architecture, 34
  big data, 42–43
  centralized system, 38

V. Trivedi, *How to Speak Tech*, https://doi.org/10.1007/978-1-4842-4324-4

Database systems (*cont.*)
  concurrency, 40
  data, 32
  data model
    NoSQL, 37
    object-oriented model, 37
    object-relational model, 38
    relational model, 35–37
    XML, 38
  distributed system, 39
  entity-relationship model, 35
  hardware, 32
  optimization, 42
  security
    data encryption, 41
    DBMS, 41
    goal, 42
    MyAppoly, 41
  software, 33
  users, 34
Debugging process
  fix bug, 76
  functionality layer, 74
  logical error/missing statement, 76
  presentation layer, 74
  regression test, 76
  reproduce the problem, 73
  tracking, 73
  unit layer, 75
  unit tests, 75
Decentralization, 141, 145
Decentralized application (dApp), 154
Deep learning, 134
Degrees of freedom (DOF), 163–164
Denial-of-service (DoS), 99
3D modelling, 160, 162, 165
Distributed database, 140
Document Object Model (DOM), 28
Dynamic HTML (DHTML), 28

E

Encryption, 102–103
Entity-relationship model, 35
Equifax data breach, 93, 107
Exchanges, 151–152

F

Factors of authentication, 103–104
Features, 132
File transfer protocol (FTP), 8
Front end technology
  Ajax, 29
  CSS, 26
  HTML, 24–26
  information design, 23
  interaction design, 23
  JavaScript
    DHTML, 28
    DOM, 28
    event handler, 28
    scripting languages, 28
  portability and accessibility
    responsive design, 30
    web standards, 30
  XHTML, 26
  XML, 26

G

GET and POST method, 4
Google, 79, 81

H

Head-mounted display (HMD), 160
Heartbleed, 107
Hosting
  bandwidth, 8
  cloud computing
    AWS, 12
    benefits, 12
    disadvantages, 13
    information and
      software, 10
    NIST, 11
    personal computer, 10
  definition, 7
  disk space allowance, 8
  FTP, 8
  prices, 9
  reliability and uptime, 8
  server type, 8
  types of, 9

HTTP requests, 89

Hybrid apps, 113–114

Hyper text markup
    language (HTML), 24

I

Immutable ledger, 144, 148

Infrastructure as a Service
    (IaaS), 12

Initial coin offerings (ICOs), 155

Integrated development
    editors (IDEs), 22

Internet
    ARPANET, 1
    components, 2
    HTTP
        MyAppoly, 4
        server, 4
        TCP/IP, 5
        URL, 4
    IP, 2
    packet switching, 2
    TCP, 2

Internet of Things (IoT), 119–127
    applications, 124
    device, 120, 123, 125–126
    platform, 120–121, 123, 125

Interoperability, 121, 127

iOS, 114–115

Iterative and incremental
    development (IID), 67

J

JavaScript
    DHTML, 28
    DOM, 28
    event handler, 28
    scripting languages, 28
    tagging, 83

JavaScript Object
    Notation (JSON), 52

K

Keyword density, 79

L

Library, 52

Lightning network, 153

Linear relationships, 134

M

Machine learning (ML), 130, 138

Mining, 149, 152

Mixed reality (MR), 158

Mobile application, 109–113, 116–117

Mobile-first development, 110

Mobile web apps, 112

Model, 133–134

Model-view-controller, 74

Moore's law, 123

MyAppoly, 4, 81

N

Native apps, 112–113

Network security, 105

Nonlinear relationships, 134

Non-relational model, 37

NoSQL, 37

O

Object query language (OQL), 38

Open-source projects
    definition, 53
    FSF definition, 53
    OSI, 54
        academic licenses, 56
        free redistribution, 54
        integrity, 55
        license distribution, 55
        modifications and derived
            works, 54
        no discrimination, fields of
            endeavor, 55
        no discrimination, persons or
            groups, 55
        reciprocal licenses, 56
        source code, 54

Open Web Application Security Project
(OWASP)
CSRF, 98
injection, 98
XSS, 98

P, Q

Packet sniffing, 84

PageRank, 79

Penetration testing, 107

Performance, 88
backend
Netflix, 88
PHP code, 88
frontend developers, 88
cache, 89
compression, 89
HTTPrequests (see HTTP requests)
minification, 90
scripts and stylesheets, 90
improve performance practices, 88
load balancer, 91
round-robin approach, 91

Phishing, 96–98, 104–105

Photogrammetry, 162

Platform as a Service (PaaS), 11

Presence, 169–171, 173

Private blockchains, 156

Programming languages
APIs, 22
applicability, 21
assembly language, 17
back end, 16
bit and byte, 17
committed community, 21
development time, 21
documentation, 21
front end, 16
high-level languages
compiled, 18
functional, 19
imperative, 18
interpreted, 18
markup, 20
object-oriented, 19
parallel programming, 19

PHP and Python, 17
scripting, 19
IDEs, 22
library and tools, 22
maintainability, 21
reliably update, 21
talent pool, 22
technical and design
parameters, 20–21

Promoting and tracking
analytics, 81
analog program, 82
clickstream data, 82
history of, 81
JavaScript tagging, 83
packet sniffing, 84
web beacons, 82–83
web log, 82
channel, 77
grocery stores, 77
organic search, 77
paid search, 77
search engine marketing, 80
performance-based advertising
model, 81
real-time hybrid auction, 80
search engine optimization
credibility, 80
Google, 80
keyword density, 79
proximity and prominence, 79
Yahoo, 78
visualization
heat maps, 84
site overlay, 84

R

Rendering, 159

Representational State Transfer (REST), 50

Revision control
benefits, 64
centralized vs. distributed systems, 63
conflict, 63
description, 62
file locking, 64
version merging, 63

Risk-driven approach, 67

S

Satoshi Nakamoto, 146, 156

Scalability, 87

Search advertising, 81

Search engine optimization
Google
bots, 78
PageRank, 79
search query matching, 79
Yahoo, 78

Search engine results page (SERP), 77

Semi-supervised learning, 135–136

Simple Object Access Protocol
(SOAP), 50

Simulator sickness, 160

Smart contracts, 154–155

Software as a Service (SaaS), 11

Software development, 59, 65, 71
agile development, 68–69
benefits, 69
debuggingprocess (see Debugging
process)
documentation and commenting, 60
iterative and incremental
development, 67
iterations, 67
rapid iterations, 67
timeboxing, 67
program architecture, 60
code reusability, 62
maintainability, 61
separation of concerns, 61
three-tiered architecture, 61
release management, 69–70
revisioncontrol (see Revision control)
semantic bugs
logical errors, 72
runtime errors, 72

syntax bugs, 72
waterfall model, 66

Standalone mobile application
development, 114–117

Standard Generalized Markup Language
(SGML), 25

Structured Query Language (SQL), 35

Supervised learning, 135

T

Tracking, 163

Training data, 135–136

U

Unsupervised learning, 136

US National Institute of Standards and
Technology (NIST), 10

V

Virtual reality (VR), 158, 165–166

Volumetric capture, 162

W

Waterfall development model, 66

Web beacons, 82–83

Web log, 82

Web Services Description Language
(WSDL) file, 51

World Wide Web Consortium (W3C), 28

X

XR, umbrella term, 158

Y, Z

Yahoo!, 78

Printed in the United States
By Bookmasters